COLTRANE'S
PLANES & AUTOMOBILES

COLTRANE'S
PLANES & AUTOMOBILES
ENGINES THAT TURNED THE WORLD

BY ROBBIE COLTRANE
AND JOHN BINIAS

**SIMON &
SCHUSTER**

IN ASSOCIATION WITH CHANNEL FOUR TELEVISION CORPORATION

First published by Simon & Schuster, 1997
An imprint of Simon & Schuster Ltd
A Viacom Company
In association with Channel Four Television Corporation

Simon & Schuster
West Garden Place
Kendal Street
London W2 2AQ

Simon & Schuster of Australia Sydney

A CIP catalogue record for this book is available from the British Library

0-684-81957-0

The television programmes on which this book is based were produced for
Channel Four Television Corporation by Idealworld Productions Limited

Design by Design/Section, Frome, Somerset, England
Printed and Bound by Bath Press, Bath, England

ROBBIE COLTRANE'S

ENCYLOPAEDIC & COMPREHENSIVE KNOWLEDGE

of

THE ENGINES

THAT CHANGED THE COURSE OF HISTORY, BROUGHT THE WORLD TO ITS CURRENT STATE

of

ADVANCEMENT

AND MADE THE TWENTIETH CENTURY WHAT IT IS TODAY

. . .

MADE ACCESSIBLE TO THE GENERAL PUBLIC FOR THE VERY FIRST TIME

. . .

by ROBBIE COLTRANE *and* JOHN BINIAS

—————— & ——————

FEATURING SELECT EXPLANATORY DIAGRAMS

BY THE SAME R. COLTRANE

ACKNOWLEDGEMENTS

*The authors would like to thank Hamish Barbour
for his contribution to this book*

CONTENTS

INTRODUCTION

Despite the great changes the feminist movement has made to our culture, sadly there's one subject that most men still hesitate to talk about in front of all but the most liberated women. It's a subject that can't safely be broached in front of people of either sex who like to think of themselves as being cultured either. (Leonardo da Vinci took top place on the short list of honourable exceptions.) That subject is engineering. Everyone presumes it's going to be boring. Well, perhaps some people are boring about trains and planes and boats and cars. But then some people are boring about sex . . .

This book will tell the story of some of the greatest developments in the history of engineering – and therefore in the history of our culture. To some people, the claim that engines have anything whatever to do with culture will seem absurd. But imagine if thousands of years from now people were digging up the remnants of our culture, just as archeologists dig up the remnants of ancient cultures today. If these archeologists of the future found one of Picasso's great paintings, or a Rembrandt, or a chunk of Michelangelo's Sistine Chapel ceiling – and at the same time they happened upon a jet engine, a steam locomotive or even a lowly two-stroke lawnmower – they surely wouldn't be less interested or impressed and intrigued by the engine than by the artwork? They'd want to know what the engine was, what it was for, how it worked and who constructed it. (Who was this mysterious warlord ATCO of Derby, they would ask one another. And what other weapons of mass destruction did he have at his disposal?) And when they'd answered those questions they'd want to know the broader implications. What did the engine enable people to do? What effects did it have on their way of life?

Now isn't it a startling thing that although our lives depend upon engines to an ever increasing extent, the majority of people can't provide even a rough answer to these questions? What would the archeologists of the future think if they discovered evidence of this phenomenon? They might imagine that engineering – the most powerful body of knowledge in human history – was controlled by a terrifying occult who used mumbo jumbo to protect their expertise from the uninitiated. Or they might simply conclude that the people who knew so little about the things on which their daily lives depended so greatly were very short on curiosity. Perhaps there's a bit of truth in both ideas.

The big story of engineering is how the power of fire was harnessed to serve human purposes. That's the story this book tells. Everyone knows that the uses that engines are put to aren't always good ones. Even so, the engines themselves are always fascinating (as, believe it or not, are some of the great engineers). Some of them are quite straightforward; some are so fiendishly clever that if you didn't know they existed you wouldn't imagine they were possible. That said, they're not so fiendishly clever that anyone capable of navigating their way across town with an A-Z won't be able to understand them. Making that understanding possible, along with telling the big story – the story that ususally gets lost in a welter of detail – is the aim of this book.

WHAT IS AN ENGINE?

A few hundred years ago the word engine meant any kind of natty contrivance, mechanical or otherwise – any old piece of ingenuity, in fact. But over the years the word has come to be used solely to refer to what used to be called heat engines. Heat engines burn fuel and convert its energy into work. It's heat engines that we're interested in. They're the engines that have really changed the world. The silicon chip has certainly changed our culture and our society, but as yet it's had little impact on the face of the planet – apart from its role in guiding and controlling the work done by heat engines. Heat engines are where the power is; they're the devices that have allowed us to cross continents, decimate rainforests, generate electricity, fly right around the globe in hours and reach the moon in days.

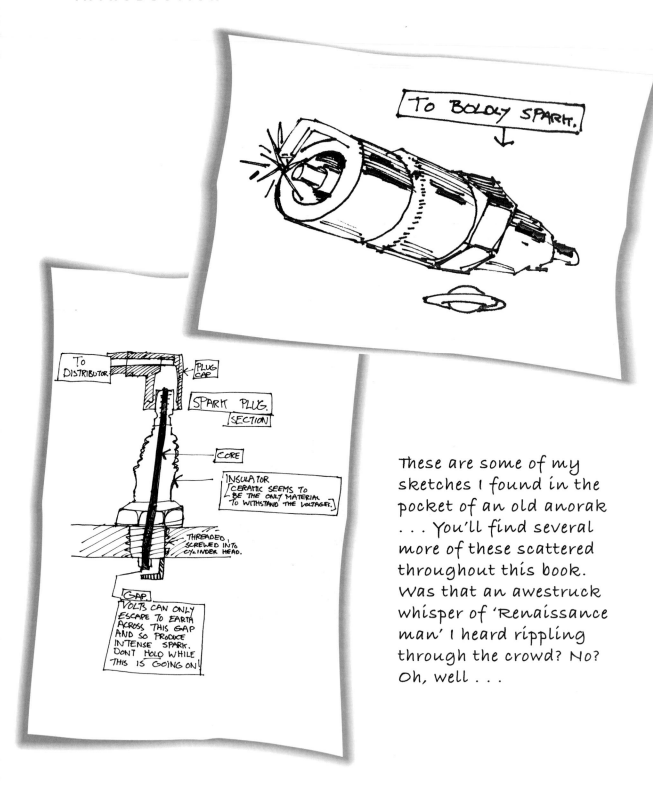

TO BOLDLY SPARK.

TO DISTRIBUTOR

PLUG CAP

SPARK PLUG.
SECTION

CORE

INSULATOR
CERAMIC SEEMS TO
BE THE ONLY MATERIAL
TO WITHSTAND THE VOLTAGE.

THREADED,
SCREWED INTO
CYLINDER HEAD.

GAP.
VOLTS CAN ONLY
ESCAPE TO EARTH
ACROSS THIS GAP
AND SO PRODUCE
INTENSE SPARK.
DON'T HOLD WHILE
THIS IS GOING ON!

These are some of my sketches I found in the pocket of an old anorak . . . You'll find several more of these scattered throughout this book. Was that an awestruck whisper of 'Renaissance man' I heard rippling through the crowd? No? Oh, well . . .

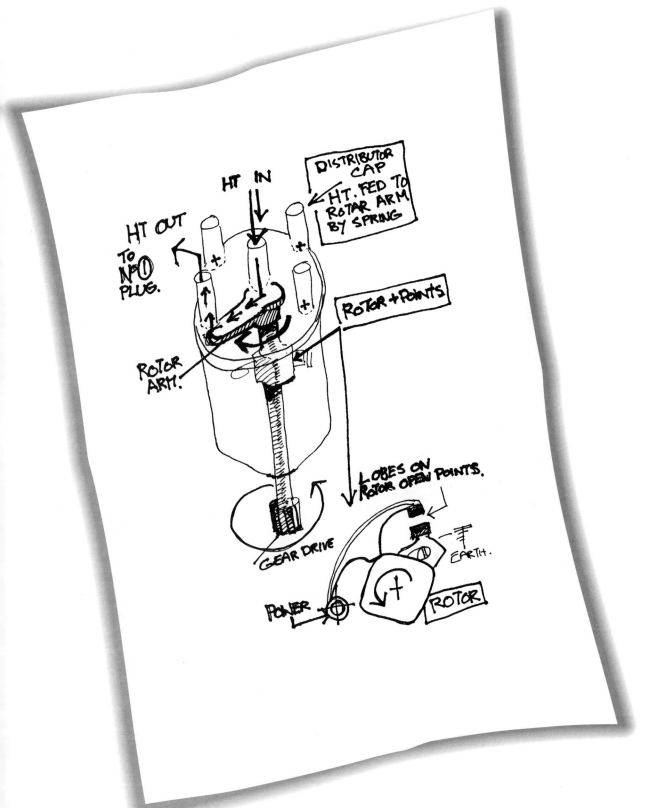

HT IN

DISTRIBUTOR CAP
←HT. FED TO ROTAR ARM BY SPRING

HT OUT TO N°① PLUG.

ROTOR + POINTS

ROTOR ARM.

LOBES ON ROTOR OPEN POINTS.

GEAR DRIVE

EARTH.

POWER

ROTOR

CHAPTER ONE

V8
THE AMERICAN ENGINE

How can an engine possibly have character? It's easy enough to see that cars have character. They have the aesthetic qualities of form and colour and line; like furniture and architecture, they have decorative details; and they have a certain build quality and a certain degree of luxury. And on top of this they have certain drive qualities – understeer or oversteer, quick or slow acceleration, a smooth or hard ride, heavy or light clutch. But an engine's just a heap of carefully shaped pieces of metal bolted together in a precisely organised way. It's purely functional – there solely to turn fuel energy into motive power. So long as they do their job properly you don't even see them.

But if that was the whole story, it would be impossible to account for the affection and interest that the V8 inspires even in people who are generally immune to the pleasures of fine engineering. So what is it that gives the V8 its character? Well, first of all it has to be the sound. There's nothing in engineering that matches the sound of a well-tuned V8. It has a kind of relaxed, not-bothered sound – a little bit like the sort of confident, soft, plummy voice that comes naturally to people who have never been pushed to their limits. And then there's the feel you get from driving a V8: there's a smoothness and an eagerness in the way a V8 revs. It's equally happy being driven hard or being driven in a leisurely manner.

Technically speaking, the V8 story is a neat one. There are many uses for the internal combustion piston engine – diesel and petrol, two- and four-stroke – for which a single cylinder is quite adequate. And there are some very clever ways of increasing the power output of an engine without increasing its size. But when all those avenues are exhausted, in order to make an engine more powerful you're obliged to increase its capacity.

You do that by enlarging the cylinders. But enlarging the cylinders means enlarging the pistons as well, and this causes

"The first V8 I purchased was an Oldsmobile Cutlass Supreme. It looked like a four-wheel panty liner, very slim and long and round, and it was the colour of a tart's handbag – red, with an opalescent green metal flake-type leatherette finish on the inside. It was a designer's nightmare, but it had a 350 Oldsmobile Rocket engine inside it and automatic transmission. Sitting in the driver's seat you had no idea at all how long the bonnet was. The bumpers and everything were obscured by the curvacious lines of the body. So when I first got it, I sellotaped garden canes to the corners of the bumpers. I don't know what people must have thought. I felt a bit of a tube but suffice to say I never hit anything and when I took the canes off I knew exactly where the bumpers were."

problems. A reciprocating internal combustion engine is not entirely dissimilar to a gun – probably the single most important difference between the two devices being that while it is quite desirable for a gun to send its bullet flying as far as possible in the same direction,

"I well remember the first time. I think we always do, don't we? I was seventeen at the time – a bit old by modern standards, but then it was pretty much par for the course. It all started on a muddy field in Canada – rugby, we called it then. Somebody tried to remove my liver with a heavily studded boot. Heroic but groggy, I was dragged off the pitch and thrown in the back of a car – I knew not what sort – until in a brief moment of sentience I caught a glimpse of the name plate. It was a Ford Galaxy 500. As we sped towards the hospital and I heard that big engine burble, I realised my life would never, ever be the same again."

for an internal combustion engine to do the same to its piston would be little short of disastrous. Once the piston reaches the bottom of its stroke it has to be turned right round and accelerated back up the cylinder. Then it has to be turned round again and accelerated back down the cylinder – and so on. But shooting a piston up and down the cylinder a hundred or even two hundred times a second causes a great deal of stress. And the larger the piston, the greater its inertia and the greater the stress it produces on the major moving parts of the engine. On top of which big pistons just won't accelerate as easily or as quickly as small pistons – and so neither will the engine which is driven by them.

There are two broad ways of tackling these problems. The first is to reduce the maximum engine speed; the second is to keep the size of the cylinders down and increase total engine capacity by adding extra cylinders. The first approach is the one adopted in the massive cathedral diesels that power ships and generate electricity. Engines of this size rotate at only 80 revolutions a minute – little more than two strokes of the piston per second. Because the massive pistons move up and down fairly slowly the stresses don't get out of hand. But an engine's power is measured by the amount of work it can do in a given time. So all else being equal an engine that turns more slowly produces work more slowly too, and is therefore less powerful. Car engines have to produce a lot of power for their weight, so in this case slowing the pistons down isn't a good solution to adopt. That's why the second way of limiting engine stress is used: keep the size of the cylinders down and multiply their number. Now all that has to be decided is how many to have, and how to arrange them.

Piston engines convert the reciprocating motion of the pistons into rotary motion by means of cranks. Add another cylinder and piston to your engine and you're obliged to add another crank as well. The power output of the cylinders is combined by connecting the cranks together on a single crankshaft. The obvious way to proceed is to put your cylinders in a line. Each time you add a cylinder you add another crank to the crankshaft. This is the in-line configuration. It's ideally suited to four-cylinder engines. And in-line sixes, though at first they faced tremendous problems with vibration because, from the point of view of a four-stroke cycle, they have an 'uneven' number of cylinders (see *Two-Stroke*), were effective and popular engines.

But if you want a bit more cubic capacity without getting into serious difficulties with the vibration, stress and sluggish acceleration caused by big reciprocating masses, you need to add a couple of extra cylinders. For a long time good in-line eights weren't possible. The materials and manufacturing processes necessary to build a crankshaft that was long enough to accommodate eight cylinders, while still being stiff and strong enough not to flex, vibrate and sometimes break, did not exist. And even when they became possible, putting eight cylinders in a line makes for a very long engine: much too long to mount sideways on and a little bit too long to be accommodated lengthways under all but the longest bonnets. It is possible to shorten an eight-cylinder engine by placing the cylinders in two horizontally opposed rows of four, connecting both rows of pistons to the same central crankshaft. This would have the advantage of allowing the reciprocating motion of the pistons to be matched side by side and end to end, giving the most well-balanced and vibration-free configuration possible – but then the engine would be too wide to be practical.

Enter the V8

The V8 engine is a compromise between the impossibly long straight eight and the impossibly wide flat eight. It consists of two

"This is more of an MG road than a Ford Galaxy road, if you've any regard for your fellow human beings. And so you can see why cars like this were never big in Scotland. Well, they were big, but not popular. It's impossible for two of them to pass on anything narrower than a motorway."

The basic Crank

"Cranks can be seen all over the place. They're common on public transport and you also find them on bicycles. Don't take this personally, but when you ride a bike your shins are like connecting rods and your knees are like pistons."

"Never in the history of engineering has a petrol engine been so perfect and yet so politically incorrect. It's big, powerful, lazy, horribly uneconomical – and yet one earful of the exhaust can drive the greenest of greens to defection. To use an engineering term, the V8 is the absolute business."

THE CRANK TURNS RECIPROCAL MOTION IE BACK-AND-FORTH, INTO CIRCULAR MOTION.

THIS DEVICE WAS INDESPENSABLE TO EVERYTHING FROM THE BICYCLE TO THE LARGEST STEAM LOCOMOTIVE

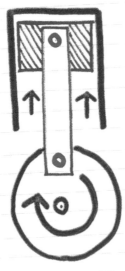

AND IS THE HEART OF ALL PISTON ENGINES.

banks of four cylinders. The banks of cylinders are located side by side and angled at ninety degrees to each other. As in an in-line eight, all eight cylinders are connected to the same crankshaft – but because the cylinders are in two rows, the crankshaft needn't be any longer than that of an in-line four-cylinder engine. So there's no problem with crankshaft stiffness or strength, and no problem with vibration. And since the crankshaft doesn't have to be as heavily built as a straight eight layout there's a weight saving as well. The whole engine is wider than a four, but considerably narrower than a horizontally opposed engine.

So technically, what the V8 amounts to is a configuration of cylinders which yields a combination of compact size, good balance, light reciprocating masses and good air/fuel distribution. Its advantages are evidenced by the fact that almost all the Formula One racing teams run V8s. But culturally the V8 is much more than a technical solution. The V8 is the American engine. It's as much an American tradition as enjoying a watery tin of beer on your front porch. It's not that other countries never built cars with V8 engines; it's just that the Americans were the only people to give you a V8 as standard in cars that were the market equivalent of the Ford Granada. In Europe, V8, stands for sporty performance in a TVR, toffee-nosed luxury in a Rolls Royce or technological brilliance in a Formula One car – while in America it's the name of a brand of tinned vegetable juice.

The first person to put a V8 engine in a car was a Frenchman by the name of Clément Ader – founder of the bizarrely named engineering company Societé Industrielle des Téléphones-Voitures Automobiles Système Ader. Ader entered seven cars for the 1903 Paris-Madrid Rally. The V8s were made up of two V4s bolted together. It would be nice to be able to say that the V8s distinguished themselves from the beginning, but at this time racing cars were rather dangerous contraptions. They were open topped, small and light with no brakes to speak of but nonetheless powered by rather large engines. No one thought it necessary to close off the roads for the purpose of racing, or to prevent the spectators (of whom there were over two million along the route) and livestock from wandering on to the road during the race. Consequently, by the time the race reached Bordeaux the death toll already stood at ten. The V8s did at least make

"V8 – the ideal combination of chocolate sticks and minirolls"

it that far. The race was stopped and the competitors were sent back to Paris by train, suffering the symbolic castration of being banned even from driving their vehicles to the railway station, being pulled there by horses instead.

The Ader V8 was followed by the Darracq. This modestly sized 24-litre V8 took the world landspeed record in 1905. It was taken to America, where it caused a bit of a stir, and was driven by Louis Chevrolet, the racing driver who gave his name to the big US car manufacturer. Unsurprisingly, the Darracq never went into production as a road car. The first V8 to do so was given what has to be the least sexy name in the history of motoring: the Rolls-Royce Legalimit, so

The eight-cylinder engine of the Legalimit

called because it was mechanically restricted to the then national speed limit of 20 m.p.h. Very few Legalimits were built. The first V8 to be manufactured in anything more than handfuls was made by the French company De Dion Bouton, who made their first production V8 in 1910. They had it running well by 1913 and kept the same model in production until 1923.

De Dion Bouton

The first indigenous American V8 was built by Cadillac in 1915. In 1910, eighty per cent of US cars had four cylinders, and Cadillacs were no exception. The four-cylinder configuration was really inadequate for the big displacements of those early, low-compression engines, but it was forced on manufacturers by the absence of ignition systems and carburettors adequate to service more than four cylinders. A few manufacturers had started to

TYPICAL STRAIGHT-8 LAYOUT:

'The inlet feed for 2/3 and 6/7 are different lengths. This leads to different quantities of mixture reaching different cylinders. The same problem exists with the exhaust. There tends to be a build-up of pressure in the manifold, which prevents burnt gas from escaping from the cylinders (the notorious 'back pressure' that bikers worry about).'

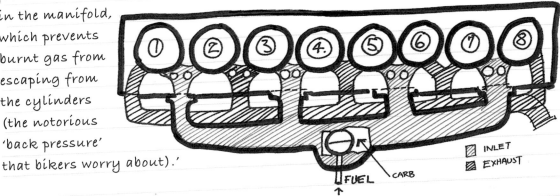

SCHEMATIC LAYOUT OF 50S V8:

'One of the incidental advantages of the V8 layout is that the carburettor - where petrol vapour gets mixed with air - can be placed in such a way that it is pretty much equidistant from all eight cylinders. This is important because every pipe gives resistance to the flow of gas along its length and the resistance is in proportion to the length of the pipe. The remaining difference in distance between the carb and the cylinders can be equalised with an ingenious bit of casting, as shown here in red.

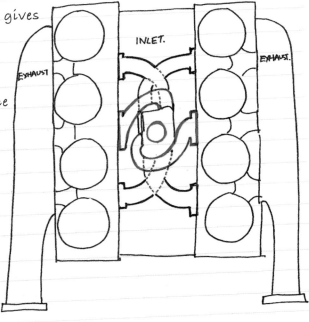

experiment with straight sixes, but early models suffered from serious vibration problems. Often the vibrations were bad enough to break the crankshaft. So Cadillac had steered clear of the troublesome six, not wishing to sully their reputation for high build quality. But when the six-cylinder pioneers overcame their development problems, Cadillac was left way behind. Too late to benefit much by developing a six of their own, they decided to go two better by developing an eight-cylinder model. The V8 configuration allowed them to equal the cubic capacity of the sixes while having lighter pistons and a lighter, shorter and stiffer crankshaft. So their engine would accelerate faster, run smoother and reach higher engine speeds. The Cadillac company bought a De Dion Bouton V8 and dissected it thoroughly to learn all they could about the V8 engine before designing a new one of their own.

The Cadillac V8 was a great success. It established the V8 as the ideal configuration of cylinders for big automobile engines, and over the next few years the V8 became the standard for luxury vehicles. And so, of course, after a while every self-respecting American wanted one. And who better to see that they got one than Henry Ford? Ford's Model T had famously made motoring a possibility for people who didn't have money to burn, and though in the late twenties he had shown where his real sympathies lay by declaring that he wasn't interested in any engine that had more cylinders than a cow has teats, it was Ford who did the same thing for the V8.

It wasn't that Ford was exactly keen to give luxury to the people. He wasn't that sort of man. His Model T was a farmer's car: its high clearances and hard springs were designed to cope with bad rural roads. But roads had improved and people were becoming more interested in style and comfort than utility. The

A custom 1915 V8 Cadillac. The engine was based on the V8 De Dion Bouton, the bodywork on D. McCall White's mother's parlour curtains

"Cadillac did this terrific publicity stunt. They took four engines apart, mixed the parts together and invited members of the audience to re-assemble the engines. They did, and all four engines ran just as well with the new combination of parts. Back in the days when engines were built individually, hand crafted and hand fitted, this uniformity of parts was a tremendous achievement."

THE CADILLAC V8

In order to gain a step on their competitors, Cadillac kept their V8 project a secret. To prevent news from leaking out via Detroit trade gossip, they employed an outsider to take charge of it – a Glaswegian engineer by the name of D. McCall White. White joined Cadillac in 1914 and by 1915 their first V8 was in the showrooms. It had a 122-inch wheelbase, but because of the shortness of the engine block, it had the same body space as a 134-inch wheelbase car powered by a straight six. It was offered in ten body styles, ranging in price between $1,975 and $3,600. Displacement was just over 5 litres, top speed between 55 and 65 m.p.h. depending on the model, and mileage was around 12 m.p.g. Cadillac already had a reputation as a high-quality motor manufacturer and with their new V8 they established themselves in a position they have kept right up to the present day. The V8 sold a massive 15,000 units annually. It's nearest competitor was the Packard twin-six – a V12 design that was even smoother than the Cadillac, but because of the additional mechanical complexity entailed by running four extra cylinders, the Packard was less reliable. Three times as many customers bought a Cadillac V8 as went for the Packard. In the twenties Packard finally dumped their V12 and produced a V8 instead. They quickly started to outsell Cadillac.

During the 1914-18 war W. O. Bentley – working as an aircraft engineer at the time – bought a Cadillac V8. He used to start it up in top gear and drive it at walking pace around the Rolls Royce works to show off its low-end torque, and wind up the Rolls Royce engineers in the process. The Cadillac V8 saw a good deal of military service. The French army bought Cadillacs as transport for their general staff; the English and Canadians used them, mainly as ambulances; and the US army bought them as a general purpose vehicle – for the general purposes of officers, of course. Wherever it was used its reputation for reliability and easy driving grew.

ancient and ugly agricultural Model T couldn't provide either of these things. By 1926 sales were dropping and Ford, who had up till then resisted all calls for change, finally recognised that the T – having sold a modest 15 million units in nineteen years – had had its day.

In May 1926 Ford shut down Model T production while a new car – the Model A – was prepared. Such was the loyalty of Ford's customers that 400,000 orders were taken before it had even been seen

"Good old Henry Ford focussed his attention on building what he knew would represent true value for money in a car – an affordable, cast iron V8 engine: the Ford flathead."

The Ford V8 was a side-valve design. This meant the cylinder heads – normally cluttered with valves and valve gear – accommodate nothing more than the spark plugs and cylinder head bolts – winning it the name 'flathead'.

"Bonny and Clyde had one of these, and so did John Dillinger – Public Enemy Number One. Dillinger actually had the brass neck to write to Henry Ford and congratulate him on his engine."

in public. The A was light for its power, though, and nippy to drive, so it did well for a while as the hundreds of thousands of loyal customers who had been waiting for the chance to buy Ford finally got the opportunity. But though it was advanced in comparison with the T, in terms of its engine it was scarcely any different. The new in-line four produced no more power than the four-cylinder engine that powered the T, and though it was beefed up a bit before going into production, it hardly represented an advance on the competition so much as a half-hearted effort to catch it up.

So in 1928, when Chevrolet brought out a new car powered by a powerful, smooth-running six-cylinder engine, Ford was in trouble. The low-cost car market had caught up with the luxury market of fifteen years earlier: the six was by then a well understood engine and could be mass produced without great difficulty. Chevy's new engine used the same size bolts as stoves of that period and gained the nickname the stovebolt six. Marketed as a six for the price of a four, it proved to be just what people were looking for. By 1931 Chevrolet had overtaken Ford in sales. Added to the effect of the 1929 Wall Street Crash and the subsequent Depression – which reduced total automobile sales in the USA from four million a year to two million in two years – the success of the stovebolt six put Henry Ford in a hole. He guessed that to keep his place as populist number one in the US automobile industry he'd have to offer his customers more luxury for their money than they could get with Chevrolet, so he took a risk and did what Cadillac had done before him, skipping the straight six and going two better with a V8.

Ford owned the upmarket Lincoln Motor Company, who produced a very successful V8. But that engine was built from many precision machined parts; the car it powered retailed at $4,600. Ford was aiming to produce a V8-powered car for $600 – only $50 more than his four-cylinder model. The four-cylinder engines that powered the T and the A were cast and machined in a single block. Ford was determined to use the same manufacturing technique with the V8.

General Motors had already shown that it could be done. But General Motors produced about 5000 monoblock V8s in a year; Ford wanted to produce 3000 a day. So the challenge was to find a way of producing monoblock V8s quickly and cheaply.

On 9 March 1932 the first Ford 'flathead' V8 drove off the assembly line. The engine had a displacement of 3622cc, giving the

car a maximum speed of 80 m.p.h. and a 0-50 time of twelve seconds – faster than most sports cars of the day. At first the new engine suffered from excessive oil consumption, piston wear and overheating. It kept a reputation for running hot for the rest of its days, but the teething problems were soon sorted out to most people's satisfaction – and to many people's delight. The engine was a success: six million flathead V8s were sold before the war, and another six million were sold after the war had ended. The flathead Ford was the first truly popular V8 car engine.

So the American dream – at least in the realm of the motor car – was realized: a car powered by the very best type of engine for quiet, fast, comfortable and reliable motoring was available to all. The Ford flathead was put to use in all sorts of ways. It powered trucks, sports cars and racing cars (Ford won the Monte Carlo Rally with one in 1936). It was adapted for use in boats too, and in the Second World War it was used in four-wheel drive vehicles, fire wagons, tracked vehicles, personnel landing craft and even tanks.

"I could not believe – I literally just could not believe – how smooth the '32 Ford V8 was to drive. And the torque! You could drive it away in top, no problem. You could drive for miles without tiring yourself out. And it was an ordinary punter's car – for tradesmen and shopowners – people who in Britain had to make do with the Austin Seven."

This is a pretty rare vehicle - a genuine 1930s salt flats racer. Before the hot-rod became a kind of automotive virility symbol for insecure adolescents, a few mechanically minded speed freaks were stripping Ford V8s of everything that didn't make them go faster and driving very, very quickly in short straight lines on the Mojave salt flats in California. The sport was given added spice by the fact that nobody except the driver in front could see anything beyond a cloud of white salt.

"I'd always had a rather snotty attitude to American cars. Most people had in those days. I thought they were a little bit huge and vulgar and unpleasant. But when I first got a ride in one, just the noise of it was the start of an insane love affair. It was twenty years before I got the money together to get one, but the love had never left me."

It is interesting to compare the American situation in the thirties with what was happening in Britain. Because Ford's Model A wasn't doing too well over here either. But whereas the Model A failed in the USA because it was insufficiently powerful to provide the relaxed, luxurious drive that the US market demanded, it failed in Britain because it was too expensive to run. The small car had already taken off in Europe. As early as 1919 Peugeot had introduced their pint-sized 628cc four-cylinder Quadrilette, and 1922 had seen the introduction of the 750cc, 7.8 horsepower Austin Seven. So while Ford was replacing the four-cylinder Model A with a big fat V8 in America, they were bringing out the drastically underpowered four-cylinder, 940cc, 8hp Model Y in Britain.

The move towards smaller vehicles in Britain had a lot to do with government taxation policy. The success of Ford's Model T in Britain

was ended overnight in 1920 when the Motor Car Act brought in an annual tax of £1 per horsepower (vehicles that were used solely for taking servants to church were subject to only half that sum). The Model T had a 22-horsepower engine and so immediately became a financial liability to its owners. Similar taxes directed at luxury cars affected the market in other parts of Europe.

But there was also a great difference in attitude to the automobile on opposing sides of the Atlantic. America is such a vast place that the automobile quickly became seen as an important utilitarian piece of equipment, while in Europe it retained its status as a rich man's novelty for much longer. In 1930 there was one car for every six people in the USA; in Britain there was one for every thirty people. And since the US Federal Interstate highways had started being built in the twenties it became common for Americans to make run-of-the-mill journeys by road that in Europe would count as a grand tour of the

Ford never managed to regain the position of dominance he had won with the Model T, but nevertheless his V8 engine changed the face of American motoring. And it led to an interesting spin-off: the hot rod. In the 1930s young men with nothing better to do used to drive out to the salt flats in the Mojave desert in California and drive very fast in a short, straight line. The flathead V8 was the engine of choice: easy to work on, cheap enough to blow up without worrying too much about it, and flexible enough to be tuned up in a big way.

continent. Manufacturers in Europe were still making open-topped cars – which suited the taste and self-perception of European motorists, who considered themselves to be indulging in a new kind of masculine outdoor pursuit – when in America fully enclosed vehicles had already become the norm.

Consistently lower fuel prices in America also had their effect. Petrol has always been an imported commodity in Europe. It has therefore always been more expensive and supplies have been less secure. When the US automobile culture was being forged fuel economy was never a consideration, which in turn meant that the cultural differences between Europe and America could have their sway. In contrast to their European counterparts, American engineers have never had any embarrassment about doing things purely because they gave people pleasure and made their lives easier. Driving with a manual gearbox in the States is considered a deeply eccentric thing to do. American engineers sat down and thought, 'Wouldn't it be easier if you didn't have to sit for hours in traffic with your foot on the clutch? Hey, why don't we make a gearbox that changes itself?' That would never occur to British engineers: 'What? Are you saying that people are too damned lazy to change their own gears? We didn't get where we are today by being too lazy to change our own gears. I think what we should introduce is a heavier clutch . . .' The Americans say, 'Hey, you can put it in any gear you like! And we'll put in a bench seat, so you can fit your little sister in to play gooseberry when you're out on a date. Let's make it really big and then we'll get everybody in there!' Most innovations in the automobile that have anything whatever to do with comfort and convenience – electric starting, electric lights, heating, automatic gearboxes – originated in the USA.

After the Second World War the divergence between American and European car culture continued apace. The flathead Ford had been the V8's first step towards ubiquity in the USA, and ten years after the war the second step followed. American car manufacturers had traditionally avoided engines that produced a lot of power for their size like poison. Reliability was far more important to their

"A thing that inspires admiration, affection, and, dare I say it, love, wherever it goes. I'm talking, as if you didn't know it, about the V8."

"Everybody should have a shot at driving a big, fruity V8, at least once in their lives, so that they can know how perfect driving a car can be."

customers than nippy performance, light weight or small size. Even as early as 1920 the average European touring car produced double the brake horsepower per litre of its US equivalent. But perhaps because of the effect of so many American GIs having seen and driven European cars during the war, or perhaps because of the influence of the car manufacturers having built high-performance aero engines to help the war effort, after the war American engineering culture became a little bit 'Europeanised' in outlook. Only a little bit, mind – and the effect in the end was completely the opposite of what you'd expect.

The big new thing was high-compression engines. There's a lot more about this in later chapters, but in short, increasing the amount of compression that the fuel/air mixture in a petrol engine receives before it is ignited increases the power output of an engine without increasing its fuel consumption. But squeezing a gas makes it hot, so to increase an engine's compression you have first to increase the temperature at which the fuel ignites, otherwise you get pre-ignition or 'knocking', which reduces the power output of the engine and causes a great deal of wear and tear. By the late forties fuels had improved a good deal, so all of a sudden it became possible to increase compression ratios without causing pre-ignition. But when the American manufacturers cottoned on to building high-compression engines with exotic mechanical extravagances like overhead camshafts, instead of down-scaling their engines to balance their enhanced power output, they thought, 'Great, add another two litres – we'll get even more power out of it!'

Once again it was Cadillac who led the way. They brought out the first high-compression, overhead valve V8 in 1949. In 1950 they set the tone for the decade in another way, bringing out the first car burdened with fins on the rear wings.

"Back then the engine was a selling point. Now it's the upholstery, the cosmetics, Gs and the GLs and the GLXs that people talk about. The interest in engines has gone. Fewer people are involved professionally, of course. And engines were comprehensible then; now they're stuffed full of electronics. You need a £30,000 toolkit to trace a fault."

And once again, the other manufacturers followed suit. This time it was Chevrolet that swept the field. Their stovebolt six had cracked Ford's stranglehold on the popular car market and topped the sales charts almost every year since. But by the fifties it was finally waning in popularity. The powerful new high-compression V8s made their straight sixes look less exciting than ever. Chevrolet needed to get away from their staid image. Young people were becoming more affluent and car makers had to appeal to young men if they were going to succeed.

Chevrolet responded by building their first-ever V8 engine. They guessed that what people wanted out of the V8 was no longer just smooth, effort-free driving, but speed; so they developed their V8

The Chevy small block V8 engine was so popular it outsold the VW Beetle car eight times over.

engine in collaboration with an experienced racing driver who kept an eye on the engine's race-tuned potential. This paid off in a big way, because the high-compression, overhead valve V8 engine they brought out at the end of 1954 – known as the Chevy small block – turned Chevrolet from a synonym for 'stuffy' into the number one young man's car of the fifties. The 1957 fuel-injected 4.6-litre version was the first US production engine to produce one horsepower per cubic inch (1 cubic inch = 16.4 cubic centimetres). It became the biggest-selling engine in the history of the motor car, having sold well over 60 million units to date.

"People went power-mad. Their road cars had the horsepower of race cars in Britain. It was a rash of wild consumerism. The austerity years were over and people were buying speed. And the thing about those incredibly fast cars is, people went to church in them. They were bourgeois cars."

The success of the Chevy small block was the defining moment in the history of the American automobile. It started a horsepower race between US manufacturers that took their cars to almost unbelievable levels of power, size and automation. By the 1950s automatic transmission had reached a stage of development which made it possible to mass produce, and a lot of the power put out by these big high-compression engines was sucked up by extremely inefficient torque converters. But even so, there was plenty left over to give

amazing straight-line performance to these enormous hunks of metal. Victory in the horsepower race went to Chrysler, who produced a limited edition model of their New Yorker – already America's fastest production vehicle – powered by their V8 Hemi. The overhead-valve, high-compression Hemi was first produced in 1951

This is a 1955 Chevy Bel Air – the first Chevy issued with a small block V8. The man who owns it bought it when he came back from his national service – and has had it ever since. He doesn't keep it because he's nostalgic; he just can't see why the hell he should sell it when it still looks good and runs so well. That's one kind of economy that lightweight European and Japanese cars will never have. He's had to rebuild the engine a couple of times along the way, of course.

in an unassuming 5.4-litre version. The 1960 300C was powered by a 6.7-litre version which took the car to a terrifying 176 m.p.h.

There's no doubt that the V8 became an American engine because it suits their roads, their character, their way of life and their fuel prices. It's difficult in Britain to argue for an engine that does eight

NASCAR

Nascar racing is a weird, all-American variety of motor racing. Legend has it that it originated in the deep South during the Prohibition years. The people who were running liquor got to enjoy avoiding being caught by the vehicle behind them so much that they took it up as a leisure pursuit. Nascar soon became big business. The rules seem pretty much to have been designed to make it into an ideal form of automotive

advertising: competition cars have to look like a production vehicle, and the engine has to be based on a production block, but beyond that anything is permitted. Car manufacturers reckoned that people who saw a car win the big race on Sunday would quite likely go out and buy one that looked just like it on Monday, so they put a lot of attention into making sure their cars were first past the post. Chevrolet's block – having been designed with racing in mind – cleaned up. Its racing success spawned a multi-billion dollar industry in 'aftermarket' parts designed to get more power out of engines. Chevrolet started it all by offering a $59 'power pack' with a four-barrel carburettor and a special inlet manifold and exhaust; today the aftermarket in 'hot' Chevy parts is worth something like $16 billion a year. This is a Chevy Lumina – which is to say it has a

"Stock car racing is the nearest I've ever come to being stuck in a multi-storey car park with the accelerator jammed on. It's cheap, cheerful, loud and smelly – and that's just my jacket."

flimsy plastic shell that's roughly shaped like one – powered by a tuned-up small block V8 that puts out over 400 hp.

Cadillac have always led the way with the V8. And when emission controls were introduced in the early seventies they were the first in with the all-American solution to the reduced power outputs of greener engines: they built the first 500 cubic-inch engine. That's 8.3 litres – enough displacement for nine Morris Minors. It was a V8, of course. And they put it in what is possibly the most gloriously tasteless car the world has ever seen – the Cadillac Eldorado. It was the politics of 'Frankly, my dear, I don't give a damn!'

miles to the gallon, but isn't the truth of the matter that the V8 is the engine we'd all put in our cars if it wasn't for the negative factors of fuel costs and pollution? And that's why the V8 represents something that's unique to the American way of life and the

"I guess they're just too environmentally unsound for the modern age. Maybe they should have wee places where all you do is drive a big American V8 car up and down for half an hour, the way you can pay so much money to go round Silverstone if you want to drive a racing car."

political philosophy of Americanism: the absolute certainty that if something is good then it should be available to everyone, whether you're a lawyer or a linesman. And if that sounds like commie talk then it's worth remembering that Che Guevara drove a Chevrolet.

CHAPTER TWO

STEAM

If you happen to be a connoisseur of fine steam locomotives you may well recognise that the loco in the picture is a 12th class North British with the 'mountain' 4-8-2 wheel arrangement. It was built in the Queens Park works in Glasgow in 1926. And if you're well up on the wonders of the world, you may recognise that the spectacular torrent in the background is Victoria Falls in northern Zimbabwe. So, you may well ask, how on earth did a sleek, Glasgow-built beauty end up out in the bush in Zimbabwe? The answer is quite a story. . .

The story starts in the Renaissance. What has the Renaissance to do with engines? Everything. 'Renaissance' means rebirth. Traditionally, the Renaissance is supposed to have been the rediscovery of the arts, literature and natural philosophy (what we now call science) after a long period of stagnation. But actually scholars had been recovering and developing the literary legacy of the ancient world throughout the Middle Ages. Those monks were right ones for their book-learning. The Renaissance was not so much a rebirth of classical models in the arts and science as a liberation from them – a rebirth of ambition and self-confidence. This new attitude had its effects on natural philosophy. The authority of Galen in medicine and Ptolemy in astronomy were no longer sacrosanct. And, most importantly for the heat engine, Aristotle's stranglehold on physics was finally loosened. (Unfortunately for ethical philosophy, the baby was thrown out with the bathwater. Aristotle's *Ethics* lost its influence for no better reason than his physics was wrong. Centuries of good science and silly moralizing followed.) The results of observation and experiment became more highly valued than what Aristotle had written on the subject. The knowledge of the properties of the universe yielded by these early experimentswas absolutely crucial in making the first successful heat engine a possibility.

"For a couple of hundred years the history of engineering was the history of the steam engine. This history isn't like the history of kings and queens: it has an internal logic, very much like the problem-solving logic you apply to fixing a car or any bit of footering that needs doing around the home."

For those of you who aren't connoisseurs, by the way, the 4-8-2 wheel arrangement (left) means that the 12th class has four small wheels at the front, eight big driving wheels in the middle and two small wheels at the back. It's called the 'mountain' configuration because it's designed to maximise traction while still being able to go round tight bends in the track.

"By the look of them you could be forgiven for thinking that the development of the steam train was driven purely by aesthetic judgement – possibly with an old vicar at the helm, giving his thoughts on the pitch of the whistle. But it was not so. The heyday of steam was possibly the most dynamic period in the history of engineering, and those lovely old engines were driven by hard cash as much as coal."

This is as simple as a steam engine ever gets. It's basically a big kettle with a fire under it that turns the water into steam – which is a gas. The steam pushes the little piston, which turns the shaft and makes the propeller go round.

"Yes, sirree, Bob. It'll only be hours before we're away. Of course, there was a slower pace of life in the days of steam. You can't just leap on to a steam-driven boat, start her up and whizz off. There's a good hour's work to do before you get steam up. But hey, people didn't have Neighbours to watch."

Who cares about steam?

Safely ensconced in a world where jet engines dominate the skies, diesel engines dominate the seas and petrol engines dominate the roads you can be forgiven for thinking that steam engines are no longer important. With the exception of the steam turbine, which is responsible for the most of our electricity generation, in a way you'd be right. But the steam engine is an incredibly important part of our heritage. It's not just that the steam engine was the motive force behind the industrial revolution – arguably the most dramatic change the human race has been through in the whole of its history – it's also the engine that taught us the engineering skills that eventually allowed other more efficient and powerful engines to be developed. The human race arrived in the modern world riding on a steam engine.

There was one line of experiments in particular which laid the groundwork for the invention of the first heat engine. These were to do with atmospheric pressure and the nature of a vacuum. On 1 February 1663 Samuel Pepys noted in his diary that King Charles II had a good laugh at the scientists of the Royal Society, 'For spending time only in weighing the air'. But, in retrospect, it's difficult to think of a more useful question to which they could have applied themselves to because the first half-decent heat engine in the history of mankind was powered by air pressure – and gas pressure of one kind or another has powered every heat engine since.

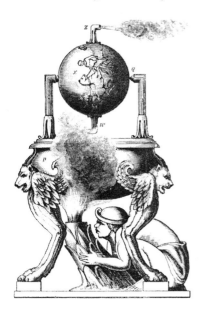

This is the steam jet as described by the ancient mathematician Hero of Alexandria. People were experimenting with steam as far back as the first century AD. But nothing came of their work – that's to say, no useful engines.

The great astronomer Galileo had once been consulted by mining engineers as to why their water pumps stopped working when they tried to raise water through a height greater than 32 feet. The grand-sounding explanation he gave them off the cuff may have satisfied them, but it didn't satisfy him. He fretted over the question for the rest of his life. It wasn't until after Galileo's death that one of his pupils, a man named Toricelli, worked out what was going on. Toricelli proved that while the longest column of water that can be raised by a vacuum is 32 feet, the longest column of mercury that can be raised in the same way is only 28 inches. Add to this the fact that 28 inches is about one-fourteenth of 32 feet, and that water weighs about one-fourteenth as much as mercury, and you have everything you need to infer the existence of a constant force or pressure which balances these columns of fluid.

That constant force is provided by the weight of atmospheric air. If gravity was switched off it wouldn't just be the tables and chairs and the tea in your cup that would float off into space; the air that surrounds the Earth would go too. It is attracted to the planet by the force of gravity, just like water is, and though it isn't heavy, there's an awful lot of it. And because air, like water, is fluid, its weight exerts considerable force in all directions. This force is called atmospheric pressure.

Despite what we say and the way we think, vacuums don't 'suck' things into them at all. When a water pump creates a vacuum at the top of a pipe by pumping out the air, what happens is that the

ROBERT BOYLE'S AIR PUMP

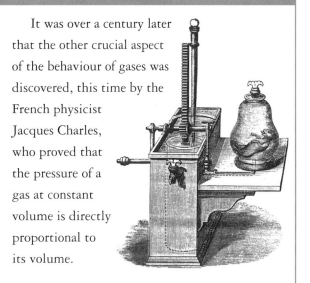

Boyle's work on the behaviour of gases under pressure was of great significance in the history of the heat engine. In 1662 he proved that at constant temperature the pressure of a gas increases as its volume increases, and vice versa. Like most enquiring minds of the seventeenth century, Boyle was fascinated by the nature of vacuums. Where the hamster comes into it I daren't even guess. But the picture's inclusion in this book demonstrates beyond doubt that if it's true that there are basically only six jokes, the hamster gag is at least two of them.

It was over a century later that the other crucial aspect of the behaviour of gases was discovered, this time by the French physicist Jacques Charles, who proved that the pressure of a gas at constant volume is directly proportional to its volume.

water is *pushed* up the pipe by the pressure of the air outside the vacuum. Atmospheric pressure isn't an easy idea to grasp, mainly because we live with it from birth till death and remain blissfully unaware of those tons of air that press down (and up, and sideways) on us. It's simple enough to understand that water pressure increases the deeper you go under the sea. Because you're not used to it, and because it increases quite quickly as you go down, water pressure is something you can actually experience. Whereas the idea that we live at the bottom of an ocean of air, where the pressure it exerts is at its very highest, is comparatively alien to us.

The first blow in the struggle to build an effective heat engine was struck in 1661 when the German scientist Otto von Guericke used the force of atmospheric pressure to drive a piston. He pumped air out of a bottle with a hand-pump and then gave the bottle to a boy who applied this bottled vacuum to a valve on an air-tight cylinder. Twenty men with ropes couldn't prevent the piston from being pushed down the cylinder by atmospheric pressure.

Another of von Guericke's experiments involved two hemispheres from which he had removed all the air. Von Guericke solemnly swore that the two hemispheres were held together by nothing but a vacuum, then cracked a rib laughing as locals fought to get their hands on the ideal Christmas present for their wives.

It's not hard to see that some mechanical advantage could be taken of atmospheric pressure. It's cheap, and unlike wind- and water-power, it's available wherever you go. The next essential step was to find a more practical means of creating a vacuum. The man who cracked this problem was the Frenchman Denis Papin. Born in 1685, Papin was a Protestant who was forced to flee his country because of religious persecution. In Britain he used the facilities of the Royal Society to carry out experiments using steam power to move a piston inside a cylinder.

Papin's cylinder was also his boiler. Fire was applied to the cylinder base, causing the water inside to boil and turn into steam. The pressure of the steam in the cylinder became greater than atmospheric pressure and the piston was forced up the cylinder. The fire was then removed from the cylinder, allowing the cylinder to cool and forcing the steam to condense back into water.

ENGINES THAT DIDN'T TURN THE WORLD

The first attempt to utilise the power of a vacuum in a heat engine was the gunpowder engine of the Dutch scientist Christiaan Huygens. Huygens knew that when gunpowder ignites it produces a large amount of hot, rapidly expanding gas. So he built a cylinder sturdy enough to withstand the force of a small explosion and ignited gunpowder inside it. The expanding gases produced by the explosion forced the piston up the cylinder. When the piston reached the top of the cylinder valves were exposed which allowed the remainder of the air and gas to escape. As the cylinder cooled, the gas remaining inside the cylinder cooled too, and so drastically decreased in pressure. The pressure inside the cylinder being lower than the pressure of the atmosphere, the piston was forced back down the cylinder. The inconvenience of operation of Huygens' gunpowder engine made it a bit of a non-starter – not to mention the complaints from the neighbours. Nevertheless, the gunpowder engine was a big influence on Huygens' research assistant, Denis Papin, who survived his exciting working environment unscathed and went on to develop the first steam-driven atmospheric engine.

There being little gas left in the cylinder, the pressure of the atmosphere pushed the piston back down the cylinder. Papin made a small model of his engine which would raise a 60-pound weight once a minute. The crude state of metalworking at the time meant Papin had great difficulty getting a piston and cylinder that fitted together closely enough to make his design practical. But the main shortcoming of Papin's engine is intrinsic to the design: the fire had to be removed from under the cylinder after each upward stroke of the piston. A less convenient arrangement can hardly be imagined.

It took a century of development in metalworking techniques to make practical exploitation of the theoretical work on atmospheric pressure a possibility. The man who took the task in hand was Thomas Newcomen, a metalworker and toolsmith, born to a fairly well-to-do family in Dartmouth, Devon on 24 February 1663. He was a devout Baptist who spent every Sunday preaching. This isn't just an unrelated snippet of biographical information, by the way. The non-conformist religions inculcated anti-authoritarian attitudes and a belief in the importance of personal industry as a means to ultimate salvation. Both were terrifically important factors in stimulating the sort of work that made the industrial revolution happen – and also explain why on a fine, sunny weekend in Glasgow, B&Q is full of guys in Rangers strips buying shelf brackets.

Newcomen had little formal education. He was apprenticed to

WHAT IS STEAM?

What you see coming out of the spout of the kettle when it boils isn't steam. Steam is a gas and, like most gases, it's invisible. What you see is the cloud of water droplets that forms when the steam is cooled down by the air and condenses. If you boiled a kettle in a room that was hot and dry enough the steam wouldn't condense and you wouldn't see anything at all – just like when you put the heater on when you're having a shower.

And if you find you can't think of steam as a gas rather than a cloud, remember that the bottles of butane gas that can be used to fuel stoves and heaters are bottles of liquid; when the valve is opened the pressure inside the bottle drops and the butane evaporates. On a planet with an atmosphere  much, much heavier than ours, butane might be found in puddles on the ground. It would evaporate on hot days, just like water does, and smoking would be illegal.

an ironmonger and blacksmith in Dartmouth, where he later set up a partnership with another Baptist, a plumber by the name of John Calley. In Newcomen's day Devon and Cornwall were peppered with tin and copper mines, and much of his work would have been related to the mining industry. One of the most pressing industrial problems of the seventeeth century was that seams of ore close to the surface were rapidly becoming exhausted. To get at new seams, mines were being sunk deeper and deeper. The deeper they got, the more flooding there was and the further the flood water had to be raised.

Some mines were being forced to close simply because their pumps couldn't keep up with the rate of flooding. Others closed because so much money had to be spent on pumping that the mine became uneconomic. Newcomen realised there was a need for a means of raising water that would be both more powerful and cheaper than the horse-driven pumps that were the limit of the current technology, and together with John Calley he set about designing a machine to do the job. He worked at the task throughout the first decade of the eighteenth century, and by 1712 he had installed a fully functioning pumping engine in a mine at Dudley in Worcestershire.

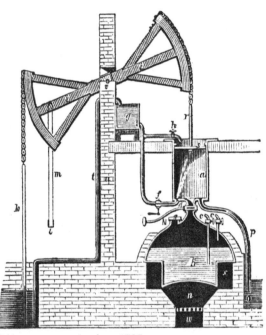

A Newcomen engine from 1705.

Newcomen's engine was elegantly simple in design. Steam was produced by a simple boiler at just enough pressure so that, when the piston was travelling up the cylinder and a valve was opened connecting the boiler to the cylinder, the cylinder would be filled with steam. The steam valve was then closed and cold water was sprayed into the cylinder. The fall in temperature caused the steam in the cylinder to condense, leaving a vacuum behind it. The pressure in the cylinder having now dropped to well below atmospheric pressure, the piston was forced back down the cylinder, pulling one end of a heavy beam down with it as it went. The beam was pivoted in the middle and the other end attached to a pumping rod. As the piston moved down the pumping rod was pulled upwards, lifting flood water with it. At the end of the power stroke, the force of

THE NEWCOMEN ENGINE

The Dudley Castle Newcomen engine worked at about 12 strokes per minute, raising 10 gallons of water through 50 yards at each stroke. The water was moved in two lifts of 25 yards each. Later versions were considerably more powerful.

The valves which controlled the admission of steam and water into the cylinder were operated by the rocking movement of the beam via a verticle 'plug rod'. One operating problem this engine was prone to was 'windlogging'. This resulted from the imperfect valve machinery; a certain amount of air entered the cylinder with each injection of steam and each injection of cold water. When air built up it was no longer possible to produce an effective vacuum by condensing the steam; the machine was windlogged. Newcomen introduced a valve to allow the release of built-up air. It was called the snifting valve because of the noise it made during operation.

The cylinder of Newcomen's engine was made of cast brass. It was not machined internally for the simple reason that no equipment capable of performing such an operation existed at that time. Instead its internal surface was smoothed by hand. The piston was also brass and, as you can imagine, not a particularly good fit. The seal between piston and cylinder was made by a leather diaphragm kept constantly moist by a few inches of water which lay on top of the piston.

Newcomen-type engines were still being used in some industries as late as 1900. The South Liberty Colliery near Bristol ran one from the middle of the eighteenth century right up until the beginning of the twentieth. Before it was decommissioned a steam enthusiast went along to check it out and found it was using steam at about ten pounds per square inch above atmospheric pressure, making ten power strokes per minute and giving a power output of around 50 horsepower.

gravity on the heavy pumping rod was sufficient to pull the steam piston back to the top of its stroke as fresh steam entered the cylinder. In other words it worked like a see-saw, with water at one end and atmospheric pressure at the other.

With the addition of a simple mechanism to open and close the valves automatically at the right moment, Newcomen had come up with the first-ever self-acting heat engine. His is far from being a household name, but if you were to choose one person to point the finger at and say, with admiration or annoyance, 'That's the man who brought us into the Age of Engines' it would have to be Thomas Newcomen. But don't bother trying to decorate Newcomen's grave with flowers – or abuse, if that's how you're disposed towards the modern world – because no one knows where he's buried.

Newcomen's engines were quickly recognised by miners to be a very good thing indeed and Newcomen, helped by his Baptist contacts, made a living from selling and installing them in the copper and tin mines of the West Country and the coal mines of the Midlands. The design was improved in small ways over the years but remained basically the same. Vast cylinders up to six feet in diameter allowed water to be pumped from prodigious depths. By the time of Newcomen's death his engine had brought the ailing mining industry in Britain back to life and was in use in many parts of continental Europe.

All this said in favour of the Newcomen engine, it was also massively and significantly inefficient. Newcomen's engine put out so little power for the amount of coal it consumed that it was only really economically viable in coal-mining areas, where fuel was relatively cheap. A typical Newcomen engine could get through thirteen tons of coal in a day. They were expensive to install, too: records show that the cost of materials alone for building one engine was over £1000 – a vast sum for the early 1700s. The bulk of the money went into the brass cylinder. Eventually, with improvements in ironworking technology, cheaper cast-iron cylinders replaced brass ones. But iron has to be cast much thicker than brass. That meant that iron cylinders took longer to cool down, so the steam took longer to condense inside the cylinder after the cold water had sprayed in, which exacerbated the engine's intrinsic inefficiency.

Newcomen's engines ruled the earth unchallenged for fifty years, until that alleged inventor of the steam engine and father of the industrial revolution James Watt appeared on the scene. Now Watt invented a lot of ingenious things, one or two of which were important. But he didn't invent the steam engine. He couldn't have, since it had already been invented. The industrial revolution would have got on quite well without him – and may even have happened more quickly: Watt was one of the foremost experts on patent law of his day and in his later years spent almost as much of his time in court stopping other people from exploiting principles which he had patented as he did working on steam engines.

Watt was born on 19 January 1736 in Greenock, a small town on

Amaze your friends and infuriate your milkman with this simple demonstration of the principle of the atmospheric engine: drop a burning match into a milk bottle and then immediately place a hard-boiled egg on top. The oxygen in the bottle is used up by the flame, a partial vacuum is created, and the force of atmospheric pressure pushes the egg into the bottle.

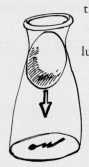

Top tip: lubricate the egg with olive oil to ease its passage.

the Scottish coast to the west of Glasgow which was a busy centre of engineering and shipbuilding expertise. Watt's father was Treasurer of Greenock and a Master Shipwright. As a child, James showed a lot of interest in his father's work and was provided with a space in Watt senior's workshop, together with a work-bench, tools and even a forge. (And I never even got a Scalextric . . .) When he was nineteen he travelled to London to work as an assistant to a mathematical instrument maker, returning to Glasgow a year later to take up an apprenticeship in an instrument-making shop. He was turned down for membership by the local guild, which caused him some problems, but fortunately someone at Glasgow University recognised the lad's promise and gave him a room from which to ply his trade, in return for doing occasional work on university equipment. It was while he was working on Glasgow University's model Newcomen engine that Watt became aware of the great inefficiency intrinsic to Newcomen's design. Out of such footering great ideas spring.

What Watt noticed was that condensing steam inside the working cylinder of the Newcomen was a waste of heat – and, since it takes coal to produce heat, a great waste of coal too. Newcomen's power stroke was produced by spraying cold water into the cylinder, reducing the temperature of the cylinder walls until they were cold enough for the steam to condense on them. All well and good – except that when the cylinder was being filled with fresh steam in preparation for the next power-stroke that steam would condense too, until enough of it had entered the cylinder to heat it back above steam's condensing point. In an engine producing ten power strokes a minute this meant the cylinder and piston were being cooled below water's condensing point, then heated back up above it again every six seconds.

Watt had a great and simple idea which allowed him to save all the energy that was being put into this fruitless cycle: he separated the cold, condensing part of the process from the hot, steam-injecting part. The condenser was kept permanently cool by a bath of cold water (which probably did its character no end of good) and the working cylinder was kept permanently hot by means of a hot steam jacket. By separating the working cylinder and condenser Watt reduced the amount of coal required to perform a given amount of work by half.

When a friend asked Watt how he was getting on with his work on the Newcomen engine's efficiency, he replied, 'You need not fash yourself any more about that, man, I have now made an engine that shall not waste a particle of steam – it shall be boiling hot.'

This was the invention that earned him his place in the history books.

Watt had it all worked out in 1765, but having the idea was the easy part. It took him three years to build a full-size working engine, and even then much more remained to be done. In 1768, he entered into a partnership with John Roebuck, owner of an iron foundry. In return for a share in the royalties, Roebuck was to help Watt develop the engine for manufacture. But Watt's design required a precision of manufacture that was beyond the standards of all but the most advanced industry of the day. The standard of craftsmanship at Roebuck's foundry was inadequate and the development work went badly.

Later the same year Watt met his ideal business partner, the Birmingham industrialist Matthew Boulton. Boulton and Watt were one of the great double-acts in the history of technology – Watt being morose to the point of despondency, while Boulton was sickeningly cheerful. Watt provided the engineering brilliance while Boulton, who ran one of the most advanced industrial operations in the world at that time, put up the capital, the manufacturing expertise and the manipulative acumen of a talented businessman. Boulton tried to buy out Roebuck's interest in the Watt patents, but Roebuck held on, and so this ideal marriage of genius and capital was not consummated for another six years. But Roebuck couldn't keep control of his patents beyond the grave and by 1774 Boulton had managed to acquire them from Roebuck's estate. Fresh development work began and in 1776 the new firm of Boulton & Watt sold its first engine.

Boulton and Watt claimed that their engine used as little as one quarter of the fuel of the 'common' Newcomen engine. This wasn't just sales talk: Boulton and Watt put their money where their mouths were, gaining much of their income from charging a royalty equal to one-third of the value of the coal saved by employing their engine instead of a Newcomen. You may have noticed that though Boulton and Watt claimed that their engine consumed as little as a quarter of the fuel of a Newcomen, Watt's separate condenser only reduced fuel consumption by a half. The extra efficiency came from the advanced manufacturing techniques of a weapons manufacturer by the name of John Wilkinson. Human beings have always gone the extra mile when it comes to killing one another, and in the eighteenth century, as it is today, weapons manufacturing was one of the highest of hi-tech industries. In order to improve the accuracy and power of his cannons, in 1774 John Wilkinson had invented a tool capable of machining the inside of a cannon barrel. This innovation could easily be put to use in the manufacture of accurately bored cylinders, which would radically reduce the leakage of steam and air round the sides of the piston, and so improve the efficiency of an engine. Wilkinson told Watt that he could bore a 72-inch cylinder that would be no more than the thickness of a sixpence out of true, and Watt jumped at the offer.

Watt's separate condenser opened up great new possibilities for the steam engine. For one thing it now became possible to use both sides of the piston to produce power – in other words, to make the steam engine double-acting. A double-acting engine produces power on every stroke of the piston – so you get twice as much power as from a single acting engine of the same size. This reduces installation costs by allowing smaller engines to do more work. It also makes the steam engine more suitable to producing rotary motion; a flywheel doesn't move as evenly or as smoothly if it is powered only half of the time. Watt achieved the double-acting engine by closing off the top of the cylinder and producing an arrangement of valves whereby steam was injected at the top and then the bottom of the cylinder in turn. As steam was being let in at one side of the cylinder the other side was being connected to the condenser. The force of atmospheric pressure acts up as well as down, of course; Watt's arrangement meant having the atmosphere push the piston both ways.

Incidentally, this innovation would not have been possible if it had not been for John Wilkinson's precision engineering. As we have already mentioned, the Newcomen engine made the seal between piston and cylinder with a band of rag, kept moist and therefore reasonably airtight by keeping a few inches of water on top of the piston. When Watt started using both sides of the piston for producing power this system was no longer available to him. He experimented with many alternatives, none of which was really satisfactory, and so it was only in 1776, when John Wilkinson's

By condensing steam in the separate condenser Watt was in effect applying a vacuum to the cylinder, just like Otto von Guericke with his bottled vacuum. The condenser is kept in a state of vacuum by means of an air pump. When the cylinder is full of steam and the piston is at the top of the cylinder a valve leading to the condenser is opened. The steam rushes out of the cylinder and into the condenser where it turns back to water. A vacuum is formed in the cylinder and atmospheric pressure pushes the piston down.

precision machining technique became available, that Watt's double-acting engine became workable. There's always a tendency to think of invention as a simple matter of ingenuity and hard work, but it was usually a combination of these two things together with some parallel technological development – in this case in metalworking techniques – that produced the big advances.

Using both sides of the piston had a massively important consequence: it gave Watt the option of using the expansive force of steam. The expansive force of steam is the force we're all familiar with – the force you see in action when a pan of boiling water pushes its lid up. Engines had hitherto used steam at around 20 psi (psi = pounds per square inch). That's just a few pounds per square inch above atmospheric pressure, which hovers at around 14 psi. Boilers, pipes and valves had improved a little since Newcomen's day and steam could now be produced safely at 25-30 psi. But until Watt closed off the top of the cylinder to make a double-acting engine there was no advantage in employing the expansive force of steam, for the following reason. Arrange to have atmospheric pressure on one side of your piston and a perfect vacuum on the other and you have 14 psi acting on your behalf. But put steam at, say, 27 psi on one side of the piston and allow atmospheric pressure (14 psi) to have its way on the other and you only have 12 psi acting on your piston. So you get reduced power and at the same time have to go to the expense of building equipment that will cope with the higher pressure, as well as using more coal to produce the extra steam. This all changed with the double-acting engine: 27 psi could now be employed against a vacuum – so giving 27 psi effective pressure. That meant Watt could increase the force of each stroke of the piston considerably while keeping the steam pressure within safe limits – a benefit that made it well worthwhile going to the trouble of producing steam at higher pressure.

So the double-acting engine with separate condenser allowed Watt to double the number of power-strokes and, at the same time, increase the force of each stroke, reducing its operating costs by 75 per cent. Great stuff. But it didn't exactly revolutionise industry. The problems were partly intrinsic to Watt's engine and partly to do with the deadly duo of Boulton and Watt. The separate condenser with its obligatory

HORSEPOWER – THE FULL STORY

It was James Watt who came up with horsepower as a unit of measurement for the power output of engines. It is said he performed experiments with dray horses in order to fix the unit at a realistic level. The horsepower is based upon the amount of work done when a horse walks in a 24-foot circle two-and-a-half times a minute, exerting a pull of 180lb. This is equivalent to 33,000 foot/pounds of work per minute – considerably more work than the best horse is capable of maintaining over a working day. Perhaps Watt was making a point: unlike horses and people, engines don't get tired. The Watt was introduced as a unit of measurement of power to replace James Watt's horsepower: 1 horsepower = 745.7 Watts. The horsepower continues to be used for engines, though. It's easy to see why it retained its popularity when physical work is being discussed – and easy to see why the term horsepower didn't catch on when it came to measuring the power output of lightbulbs and electric fires.

Brake horsepower (bhp) is the term used when the power output is actually measured at the engine shaft. With larger engines it is difficult to make such measurements, so indicated horsepower (ihp) is used instead. Indicated horsepower is a theoretical calculation of power output, based on a measurement of the cylinder pressure. An engine's power in brake horsepower is equal to its indicated horsepower minus engine friction – which is usually around 10 per cent of indicated horsepower. So bhp is ususally approximately 90 per cent of ihp.

air pump, combined with the double-acting piston, went together to make quite a complex bit of kit. As well as costing twice as much as an equivalent Newcomen, a Watt engine took a lot longer to install – up to a year, in fact. The Boulton & Watt factory in Soho, Birmingham only manufactured nozzles and valves for the engines. The cylinders were machined in Wales by John Wilkinson. And although Boulton & Watt oversaw the process of manufacture and installation, the purchasers had to find the rest of the parts themselves.

('That wrought iron gate you did for me last year was great stuff, Jeremiah. I wonder when you've got a minute would you mind knocking me up a water-jacketed condenser with an air pump?' 'No bother, Bob. Large, medium or small?')

They also had to hire and pay for the craftsmen who did the installation work. This was a major headache, and a big disincentive to go Boulton & Watt.

MINUTES OF THE PARLIAMENTARY COMMITTEE ON WATT'S ENGINE BILL, 1775

Committee: *In what respect is Mr Watt's engine better than the common fire engine?*

Boulton: *The best common fire engine that I have examined has required from 3 to 4 times the coal that Mr Watt's does to do the same work in the same time – I mean to raise the same quantity of water the same height.*

Committee: *How do you account for this different effect – from the construction of those different engines?*

Boulton: *In the common engine the cylinder is robbed of a great quantity of its heat at every stroke of the piston by the following causes: firstly by the great quantity of cold water that is injected into the cylinder to condense the steam; secondly by a small column of water that lies upon the top of the piston in order to keep it air tight; and thirdly by the cylinder itself being exposed to the common atmosphere.*

Committee: *Is this engine differently formed from any other engine?*

Boulton: *Yes.*

Committee: *Has it the advantage of saving 3 or 4 times the quantity of coals?*

Boulton: *Yes.*

Remarkably, Boulton stood up to this savage display of cross-examination technique and went on to win the day. The prize: a seventeen-year extension of Watt's patent, granted on the grounds that much development work remained to be done but would remain undone if the patent was not extended. The first Boulton & Watt engine was sold the following year. Not such a long development period after all, then . . .

SUN & PLANET.

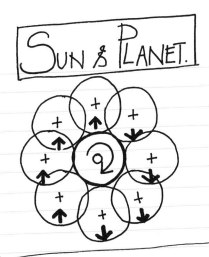

Watt's machines used a very imaginative but rather troublesome method of converting the reciprocating motion of the piston into rotary motion. It was called the sun and planet gear. It was one of five methods of producing rotary motion that Watt patented, but the only one he used.

THE PLANET, AS ALL POST-GALILEO CHILDREN KNOW, GOES ROUND THE SUN. THE PLANET DOES NOT TURN ON ITS CENTRE BUT IS FIXED RIGIDLY TO THE PUSHROD.
THE PLANET ACTS LIKE A CRANK BY TURNING THE RECIPROCAL MOVEMENT OF THE BEAM [AND PISTON] INTO CIRCULAR MOTION.

LAYOUT

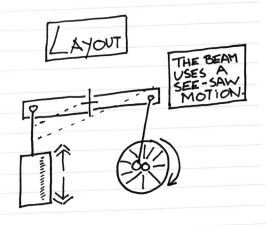

THE BEAM USES A SEE-SAW MOTION.

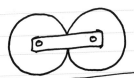

THE SUN + PLANET ARE HELD IN MESH BY A BAR WHICH ROTATES WITH PLANET.

And the complexity of Watt's engine led to maintenance problems: it was no more or less reliable than Newcomen's but a hell of a lot more difficult to fix. So while an improved steam engine could in theory recoup the extra cost of manufacture and installation in a single year through reduced fuel bills, it also tended to lead to long periods of down-time while the frustrated factory owners cast around for someone with the expertise to fix it. Watt's engine also cost more in lubricants and required more day-to-day attention by a better-trained mechanic. So

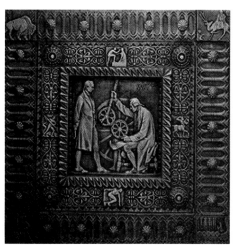

A bronze relief of Watt with his sun and planet gear is depicted on the doors of the American Academy of Science.

outside of Cornwall and London, where coal costs were really high, a Boulton & Watt engine was hardly worth the trouble. Watt himself admitted that 'When Engines of our Construction are not carefully attended to and kept in order they may burn more coals than a Common Engine that is well kept . . .' (Note that use of 'common' to refer to the Newcomen. Makes you want to fit one to run your central heating . . .)

As for replacing water power as the prime-mover in industry, the Watt engine did nothing of the kind. Even when it was perfected, the double-acting engine, when used to produce rotary motion, was still capable of suddenly reversing the direction of the fly-wheel – and so of all the machinery in the factory. In cotton mills this could lead to astronomically expensive damage to delicate weaving machinery. And Watt's engine, though considerably smoother than Newcomen's, was still nowhere near as smooth as the waterwheel, which gives a very constant rotary motion and was the first choice for quality cotton production right up until the 1850s.

So, all in all, despite its continuing fame, Watt's engine didn't make that much difference. What undeniably did make a difference was Watt's patent. Indeed, a large part of Watt's historical reputation as an engineer can be put down to the fact that no one in Britain knew more about patent law, or was better able to manipulate it, than James Watt. He specified his patents in terms of principles rather than designs, so as to prevent slightly changed copies making holes in his royalty cheques. This meant that Watt's patents covered several independent innovations and prevented them from being marketed

PATENTS

The grant of monopoly rights over the commercial exploitation of an invention is a time-honoured practice. In Britain it was originally part and parcel of the Monarch's prerogative to grant monopolies to any individual or company in return for financial or political advantages. But the monarch's self-interested exercise of this lucrative right was not to everyone's taste and the growing 'free enterprise' lobby fought long and hard to have the procedure formalised. In 1601 Elizabeth I agreed that no further monopolies would be granted without a legal trial to ensure that the good of the people was being considered. In 1624, during the reign of James I, Parliament passed the Statute of Monopolies. Monopolies other than patents of invention were abolished. New patents were issued for a fourteen year period. This was twice the period of apprenticeship in that day, so fourteen years was judged an adequate period to train workmen in the manufacture of the new product. The law stood in pretty much the same way in 1775 when Watt managed to wangle a seventeen-year extension of his 1769 patent.

The cost of obtaining a patent was prohibitive – £100 for England alone in the early nineteenth century. The procedure was slow and obscure. Successive Acts of Parliament reduced the cost, established a period of public consultation and introduced the requirement of novelty. In 1902 patent applications were required to be examined to make sure the patent had not been anticipated and in 1919 the period of patents was extended to the current sixteen years.

by their rightful inventors. By the 1890s Boulton and Watt were spending most of their time in court, 'defending' their patents against anyone who dared to use the expansive force of steam. Watt himself was almost paranoid about high-pressure steam: when Richard Trevithick built the first successful high-pressure engine after the lapse of Watt's patents, Watt told him that he 'deserved hanging' for putting the public at such a risk – and sought an act of Parliament to ban its use. Fortunately he was unsuccessful. But he was successful in court against most of his foes, who usually gave up developing their engines as a direct result of Watt's interference. Watt's success was partly due to his skill as a lawyer, partly due to his fame – which increased the authority of his testimony as an engineer – and partly due to the fact that his other half, Matthew Boulton, was a personal friend of many of the judges who heard their cases. Do I detect traces of a funny handshake?

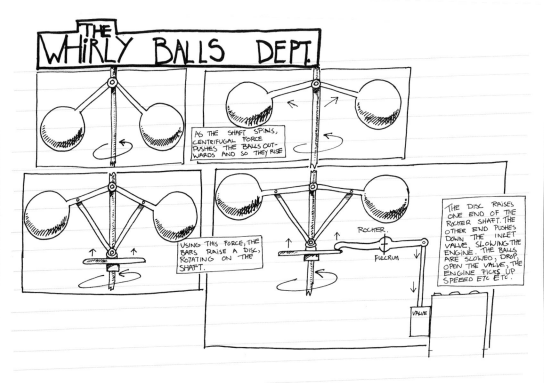

THE WHIRLY BALLS DEPT.

As the shaft spins, centrifugal force pushes the balls outwards and so they rise

Using this force, the bars raise a disc, rotating on the shaft.

ROCKER.

FULCRUM

VALVE

The disc raises one end of the rocker shaft. The other end pushes down the inlet valve, slowing the engine. The balls are slowed, drop, open the valve, the engine picks up speeed etc etc.

Another elegant Watt innovation was the fly-ball governor, which uses centrifugal force to control the speed of the engine. It had been around for ages in windmills, but Watt was the first person to use one on a steam engine – and, miraculously, got a patent out of it. The governor is set for the speed the operator desires. If the machine speeds up then centrifugal force pushes the balls further out, lifting a collar on the shaft which in turn reduces the amount of steam entering the cylinder. If the engine slows, the balls fall inward, lowering the collar and increasing the ingress of steam; an effective feedback system that massively reduces the amount of tending needed to keep the engine at a constant speed.

Watt's patent finally expired in 1800. A flurry of suppressed innovation followed. Watt had patented the principle of the high-pressure engine – that is, using steam expansively – but had been too wary of the potential dangers and too busy in courts of law, to do anything about exploiting its great possibilities. Now the men with the courage to do so were given free rein.

The best of them was Richard Trevithick, one of the great eccentric geniuses of engineering history. In 1810, while he was using one of his patent iron tanks to raise a sunken wreck off Margate, Trevithick had succeeded in raising the ship, which was being towed into shallow waters, when a dispute regarding payment broke out with the owners of the vessel. In a grand gesture of contempt that must have delighted his hard-up wife and family, Trevithick made an end to all argument by cutting loose the lashings and allowing the newly raised ship to sink back to the bottom of the sea.

In another vivid anecdote, a boy who worked in a mine where Trevithick had installed one of his engines describes Trevithick's behaviour when the machine refused to start:

Richard Trevithick as a young man. 6 foot 2 inches, good looking, and he built the first steam locomotive in history. Bastard.

> *After a bit Captain Dick threw himself down upon the floor of the engine-house, and there he lay upon his back; then up he jumped, and snatched a sledge-hammer out of the hands of a man who was driving in a wedge, and lashed it home in a minute. There never was a man could use a sledge like Captain Dick; he was as strong as a bull. Then, he picked up a spanner and unscrewed something, and off she went...*

There are a lot of stories about Trevithick's tremendous strength. He was born in Cornwall in 1771, when people were generally much shorter than today (think of all those tiny doors and low ceilings in old cottages). And in Cornwall in particular, most men were short and stocky. So at six foot two, Trevithick must have seemed like a giant. Legend has it he could write his name on a beam six feet from the floor with a hundredweight hanging from his thumb. And at a shareholders' meeting at the Dolcoath mine, of which his father was 'captain' or manager, he once picked up a fellow by the name of Hodge, turned him about by the waist and succeeded in leaving his bootprints on the ceiling. I imagine Hodge must have found that very amusing indeed.

As a child, the village schoolmaster described Trevithick as 'a slow, obstinate, spoiled boy, frequently absent and very inattentive.' (I don't suppose Trevithick cared much for the schoolmaster either.) He excelled at mathematics, but was punished for getting the right answers by the wrong methods. His father being a mine engineer, he gained early experience of both Newcomen and Watt engines, learning the fundamentals of steam engineering so quickly that by the age of nineteen he was working as a consulting engineer.

Watt was a dirty word in Cornwall towards the end of the eighteenth century. Cornwall was one of the main markets for Boulton & Watt engines. As mentioned above, Boulton & Watt not only sold their engines at a profit, they also charged a royalty of one-third of the value of coal saved by using their machine as against a Newcomen of equivalent power. In Cornwall the balance of opinion was that this was a bit of a liberty. Cornish engineers were desperate to find an engine that was as efficient as Watt's which didn't infringe his patent on the separate condenser. Unfortunately for them, Watt also had a patent on the expansive use of steam and several very promising attempts to develop high-pressure steam engines had been scotched by Watt in the 1790s. So the avenues left open for Cornish engineers to explore were pretty limited.

Trevithick started experimenting with high-pressure steam in the late 1790s. He built a static model and then a model locomotive. Both worked efficiently and effectively without the use of a condenser, separate or otherwise. Watt needed to use a condenser because though he was using steam expansively he used it only at low pressures. Without taking the trouble to create a vacuum on the other side of the cylinder (and a separate condenser with its accompanying air-pump is a lot of trouble, both to manufacture and maintain) there would have been even less effective force on the piston than in a Newcomen engine.

By dispensing with the condenser and its air-pump Trevithick at once made his engine simpler, lighter, more compact, more reliable and capable of turning much faster. And because they used the

TREVITHICK'S 'PUFFERS'

The scientist Davies Gilbert, Trevithick's friend, advisor and fellow Cornishman, recalled Trevithick's scheme to use high-pressure steam expansively.

'Trevithick came to me and inquired with great eagerness as to what I apprehended would be the loss of power in working an engine by the force of steam, raised to the pressure of several atmospheres but instead of condensing it to let the steam escape. I of course answered that the loss of power would be one atmosphere . . . I never saw a man more delighted, and I believe that within a month several puffers were in actual work.'

Trevithick's engines were called puffers because of the noise they made; once the steam had been used to drive the piston down the cylinder it was released into the atmosphere, so every stroke of the piston was accompanied by a puff of steam.

expansive power of steam at high pressures his engines were capable of a much greater power-output for their size than the best Watt engines. But, for just this reason, he wasn't able to install his first full-size high-pressure engine until 1800, the year Watt's patents expired.

Trevithick's high pressure engines were a success from the very beginning. By 1804 he had built and sold more than fifty of them. Trevithick's steam engine was the first to find employment in a genuinely wide variety of industrial applications. They were used for water-pumping, of course, and for lifting ore from the mines. They were also used to power threshing machines, sugar mills, ploughs, dredgers, drills, grain-millers and in many kinds of factory. Trevithick prided himself that his engines were three times as economical as the best Watt engines.

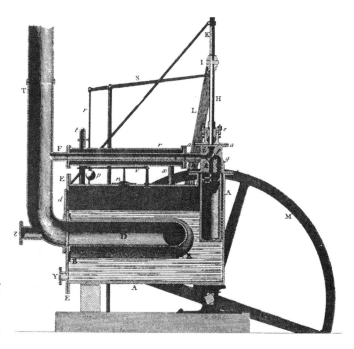

Was Trevithick's engine dangerous, as Watt had claimed it was? Well, in 1803 one of his boilers burst, killing four people. Debris was scattered over an area of 200 yards. The boiler was providing steam for a high-pressure pumping engine that was in use in Greenwich to drain the foundations of a partly-constructed corn mill. So yes, high-pressure steam was dangerous – and continues to be dangerous to this day – but only if handled carelessly. Trevithick wrote to Davies Gilbert:

Trevithick's first high-pressure engine was fitted with a condenser, and still used the rocking beam method for transmitting the drive; later engines dispensed with both these mechanisms.

It appears the boy that had the care of the engine was gone to catch eels in the foundation of the building, and left the care of it to one of the labourers; this labourer saw the engine working much faster than usual, stopped it without taking off a spanner which fastened down the steam lever, and a short time after being idle it burst. It killed 3 on the spot and one other is since dead of his wounds. The boy returned at the instant and was going to take off the trigg from the valve. He was hurt, but is now on recovery; he had left the engine about an hour

. . . I believe that Mr B. & Watt is apt to do me every injury in their power for they have done their utmost to report the explosion both in the newspapers and private letters very different to what it really is.

Trevithick wasn't intimidated by the unkind attentions of the grand old man of steam. Instead he went on to design safety devices to prevent the same thing happening again.

One hundred and fifty years after the experiments that pointed the way, steam power had arrived. Richard Trevithick had 'cast off the dead hand of Watt' and built a compact, lightweight, simple, safe, powerful and economical engine that could be adapted to a vast array of purposes – including transportation.

Trevithick was not the first person to produce a moving steam engine. A Frenchman by the name of Nicolas Cugnot had built a steam-propelled three-wheel carriage in 1769. It travelled at 2-3 m.p.h. and, once it got going, was literally unstoppable. Cugnot drove it into a wall and landed up behind bars. He tried again in 1770 but nothing came of his experiments. In Trevithick's first steam-powered vehicle, on the other hand, we have the direct forbear of the steam locomotives that were to dominate land transportation world-wide for the next 150 years. Never a man to hang around when he had a good idea, Trevithick built his first full-size locomotive in 1801.

Intended to run on ordinary roads, it was what we would now call a steam car. One of the most startling experiments Trevithick performed before building it was to push a horse-drawn vehicle along the roads around Camborne in Cornwall – startling not because of its danger or ingenuity, but because of its banality. Trevithick pushed the cart by turning its wheels. This was because before he set out to build a locomotive he first had to find out whether or not smooth wheels would be capable of providing sufficient traction to propel the vehicle. Since up to this time every vehicle in the history of transport had been either pushed or pulled, a variety of opinions existed. The most commonly held view was that some sort of toothed driving wheel would be necessary to transmit adequate driving force. Many of the earliest railways were built on this troublesome principle. By pushing a cart around by its wheels for a couple of days Trevithick worked out the right answer.

Trevithick's first run was made on Christmas Eve 1801. It worked quite well, successfully carrying several passengers up a hill. But on 28 December, no doubt under the influence of Christmas spirits, Trevithick's cousin Andrew Vivian pranged it. Davies Gilbert relates the events that followed:

> *The carriage was forced under some shelter, and the Parties adjourned to the Hotel, & comforted their hearts with a roast goose and proper drinks, when, forgetful of the engine, its water boiled away, the iron became red hot, and nothing that was combustible remained either of the engine or the house.*

As usual, Trevithick bounced right back. In 1803 he built another road locomotive and on 13 February 1804 he ran the first-ever successful railway locomotive. This vehicle was built in response to a bet. Trevithick had sold a share in the patents on his locomotives to one Samuel Homfray, ironmaster of Pen-y-daren in South Wales. Homfray made a bet of 500 guineas with Anthony Hill, a neighbouring ironmaster, that one of Trevithick's steam locomotives could haul ten tons of iron the length of the nine and three-quarter mile plateway from Pen-y-daren to Abercynon. Homfray won his bet: Trevithick's loco carried ten tons of iron on five wagons – together with seventy men – the

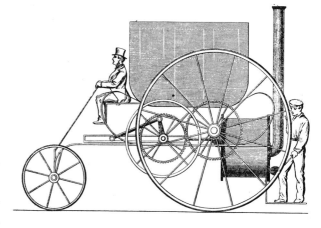

An early Trevithick 'steam car'.

whole distance in just over four hours. Four hours might sound like rather a long time, but they were slowed down by having to stop to cut down trees and remove rocks from the tracks. The engine ran at 5 m.p.h. and the boiler didn't need refilling for the duration of the journey. The only major problem was that the locomotive was too heavy for the plateway – a kind of road made of iron plates with a lip on its outside edges to keep vehicles from rolling off them. Designed for horse-drawn wagons, the plates were not strong enough for Trevithick's heavy locomotive, which caused them to crack.

After the commercial failure of his next project, a locomotive

Precursor to London Underground's cirle line? Trevithick's *Catch Me Who Can* was probably more reliable, but it was no more economically successful. The rubbernecks were charged a shilling a go, which was quite a bit of money in them days. Even so, Trevithick failed to make a profit.

A bet was placed that Trevithick's locomotive could travel farther in 24 hours than a race horse. But a rail broke and *Catch Me Who Can* overturned.

known as *Catch Me Who Can*, Trevithick decided there was no money in railways and turned his attention elsewhere. At the time, he was right. Even by 1825 only eighteen steam locomotives were doing useful work in the whole of Britain, all of them in private industry. That was the year when work began on the first passenger railway – the Liverpool & Manchester line. And even then, the directors of the railway couldn't decide whether to pull their trains with horses, steam locomotives, or ropes drawn by static steam winding engines. Trusting souls, they made their minds up after listening to representations from Robert Stephenson concerning the inefficiency of static engines and horses compared with steam locomotives. In 1829 they organised their famous competition, offering a £500 prize for the best steam locomotive. There were only five entries – and one of them was withdrawn at the last minute, incredibly because it turned out to not to be a steam locomotive at all, but rather a contraption powered by two horses on treadmills! The competition was won by George Stephenson's *Rocket*, which achieved a top speed of 30 m.p.h. Seven more locomotives of the same type were commissioned.

George Stephenson was one of the great consolidators in the history of technology. He didn't add much that was new to the steam locomotive, but he was a genius when it came to combining and adapting the best of existing designs. With the *Rocket*, Stephenson had pretty much put the whole package together. The only step that remained to be taken between the *Rocket* and the fully evolved steam locomotive was to place the cylinders

THE WHO'S WHO OF STEPHENSONS

George Stephenson (1781-1848) was born near Newcastle-upon-Tyne. His father, Robert, worked for a colliery, and George followed in his father's footsteps. By the age of seventeen he was in charge of a pumping engine. By twenty-one, he was in charge of the engines at a number of pits and was acting as an engineering consultant to other pit owners as well. He built thirty-nine steam engines during this period and replaced horse-drawn sleds at the pits where he worked with horse-drawn railway trains. In 1815 he invented a miners' safety lamp, very similar to the one invented more or less simultaneously by Sir Humphry Davy. This brought controversy, as many 'learned' people insisted that an ill-educated man like George Stephenson could hardly have achieved the same technical feat as a highly respected scientist like Sir Humphry. They accused Stephenson of copying, though there was not a shred of evidence against him.

Around this time the rising cost of horse fodder was encouraging colliery owners to look for alternative means of shifting coal, and Stephenson was commissioned to build a steam locomotive. In 1814 the *Blucher* – complete with flanged wheels – was put to work. This was followed by the *Wellington*, the *My Lord*, the *Locomotion* – built for the Stockton–Darlington railway – and then, in 1930, the *Rocket*. George Stephenson was a decent sort of man, and despite all he had to gain, he did his best to calm the manic investment in railways that was taking place at that time, in which many people lost a lot of money.

Stephenson's *Rocket*, 1829.

Robert Stephenson (1803 – 1859) was George Stephenson's son, and was brought up by his father alone after his mother and sister both died when he was young. His father took great care with his education and sent him to Edinburgh University for a year. Robert Stephenson founded Robert Stephenson & Co. in 1823, in order to build locomotives. Father and son worked very closely together for many years, but Robert eventually got an identity of his own outside of the steam locomotive industry. He built the London to Birmingham Railway, and he also designed bridges. His first bridge collapsed under the weight of a train, but his next bridge – a spectacular iron construction consisting of two vast 459-foot spans across the Menai Strait, the strip of sea that separates Wales and Anglesey – was a success. The longest wrought-iron span up until then was thirty-one feet. Robert Stephenson later became MP for Whitby and President of the Institute of Civil Engineers. He received an honorary degree from Oxford and was buried in Westminster Cathedral.

horizontally underneath the smoke box. Most of the other modifications that took place between the Rocket and the death of steam were a matter of adapting this basic design to the demand for greater power.

In the 1830s, steam locomotion took off and a massive new industry was born. Robert Stephenson & Co. exported locomotives to Germany, Belgium and France. By 1840 these countries had their own locomotive industries. Britain's early dominance in the field had its effect, though, as the English standard gauge of four foot eight and a half inches became common throughout western Europe, as well as in much of the USA and Canada. Between Stephenson's *Rocket* and the worldwide death of steam a century and a half later about three quarters of a million locomotives were built – to something like 40,000 different designs.

The effect of the railways, even on a country as small as the United Kingdom, was enormous. Until the railway network was developed in Britain there was no such thing as a national newspaper. The time taken to deliver such a publication the length and breadth of Britain would have rendered its claim to contain the news absurd. Elsewhere, steam locomotives opened up the great land masses of the world to internal and international trade. By 1850 there were 23,000

The model is Stephenson's *Rocket*. The big loco in the background is a Beyer Garratt 15th class.

STAGES OF STEAM: ONE

The first locomotive: Richard Trevithick's Pen-y-Daren engine

The basic elements are in place: A high-pressure engine with steam exhausting up the chimney providing extra draught to draw the fire forwards along the length of the boiler.

STAGES OF STEAM: TWO

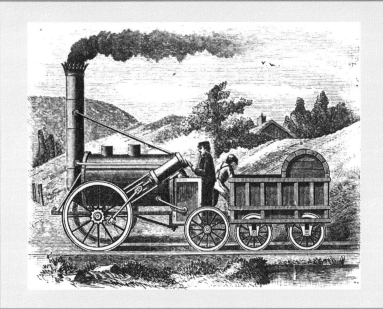

Stephenson's *Rocket*

Almost there: this 30 m.p.h. locomotive had direct drive from the piston to the wheels; flanged wheels rolling on cast-iron tracks; a multi-tube boiler and forced draft. The *Rocket* weighed 4.5 tons.

STAGES OF STEAM: THREE

The 12th Class North British locomotive 190

A classic steam locomotive, one of the most popular and successful classes from the North British Company. The cylinders are tucked away beneath the smoke box, a configuration first seen on Stephenson's Planet. The boiler produced superheated steam at a working pressure of 120 lb per square inch with a maximum pressure of 190 psi. And – since you asked – the two 20 x 26-inch cylinders coupled to the 4 foot 3-inch diameter driving wheels produced a tractive effort of 32,940lb at 75 per cent boiler pressure. The engine weighs a little over 78 tons. The tender holds 10 tons of coal and 4250 gallons of water. The Rhodesia Railways paid £8,972 for it, as part of an initial batch of twenty. Twenty-five more were to follow.

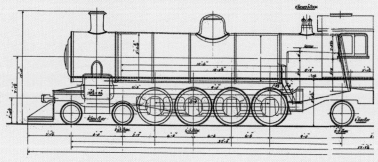

"Although it looks complicated, the steam locomotive is really quite simple."

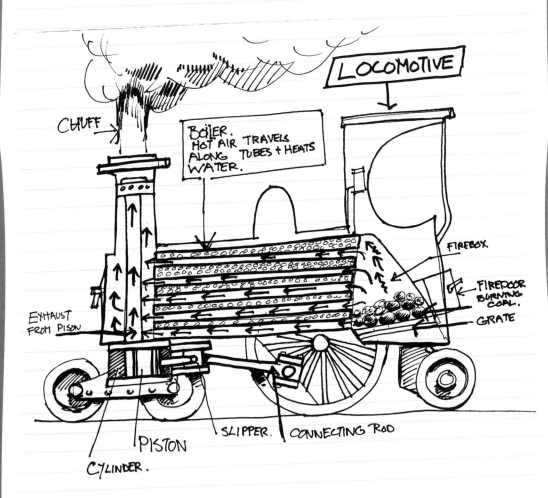

LOCOMOTIVE

CHUFF

BOILER. HOT AIR TRAVELS ALONG TUBES + HEATS WATER.

FIREBOX

FIREDOOR BURNING COAL.

GRATE

EXHAUST FROM PISTON

PISTON

CYLINDER.

SLIPPER.

CONNECTING ROD

miles of railway track world-wide – 6,600 miles of which were in the British Isles. By 1890 the figure had risen to 370,000 – a measly 20,000 of which were in Britain. Much of the rest of this track was

laid in Europe – 25,000 miles of it in Germany alone. But the USA accounted for almost half of the total, with a massive 166,000 miles of track. Starting on the eastern seaboard, railways penetrated the continent of North America until they reached the West Coast, bringing with them the internal trading conditions that made the United States the political and economic phenom-enon it is today. Although the idea of the railways was imported by the USA from Britain, the Americans very much took hold of it and made it their own, and the development of US steam locomotives took an independent course from an early stage.

Few parts of the world were unaffected by steam locomotion. The British Empire – which was, more than anything else, a trading empire – took its new invention to its colonies and dominions, reclined on the veranda with a G&T and looked on with a contented smile as goods and commodities arrived at the sea-ports by rail, ready

to be shipped out to Britain and other markets. In Africa, Cecil Rhodes planned a railway line that was to cut a line through the continent from Cape Town to Cairo. Trade, Rhodes predicted, would be picked up along the

"Despite their Jurassic appearance, these trains are actually not that old."

"I declare this piston rod connected . . ."

way. His project never got quite that far – though it was still being considered long after his death. But it did get as far as what was then Rhodesia. The first train into Bulawayo arrived in 1896. By 1904 the line had reached the Victoria Falls, and in 1905 the Vic Falls bridge was opened to traffic for the first time.

This is what you can do with the triple expansion steam engine: you can drag these monsters all the way from Manchester to dominate another continent.

Built in Glasgow in April 1926, Loco 190 was delivered to the port of Beira in Mozambique and transported by train to Gwelo, east of Bulawayo, where she was assembled. She was in active service with Rhodesia Railways by December of the same year (and now has a magnificent 2,000,000 miles on the clock). An impressively swift and efficient delivery – which brings us round to the little matter of steam navigation.

Steam engines were successfully used to power boats quite a while before railways became a practical commercial proposition. The first regular commercial steamboat service in Europe was started on the Clyde in 1812. The *Comet* was built by Henry Bell in 1811 and went into service a full thirteen years before the opening of the first public railway. She plied the river between Greenock and Helensborough and turned a healthy profit. Steam navigation was successful even earlier in the USA, where the many wide, smooth-flowing rivers made paddle steamers powered by big, heavy engines quite practical. But as you can see from the graph, ton for ton, sailing ships continued to outnumber steamships until right into the 1880s. What was holding the merchant ship owners back? Could it have been a romantic attachment to sail?

Without delving into statistics about Valentine's cards and candle-lit suppers, I think it's safe to say that when it came to calculating their

SS SHIELDHALL

"The SS Shieldhall is the largest triple-expansion screw-driven coaster in the world. It was great big filthy buggers like this that really got the Empire on the move."

The Shieldhall was built in 1955 to a 1928 design. Glasgow corporation commissioned a couple of modern vessels but had nothing but trouble from them and so reverted to a traditional design next time round. She was used to transport sewage from Glasgow down to the mouth of the Clyde. Bewilderingly, she also carried passengers. When she reached the mouth of the Clyde, a bell was rung. The passengers assembled in the tea room. The doors and windows were closed, tea was served, and the sewage was dumped. Meanwhile, inhabitants of the towns and villages on the banks of the Clyde toiled away with pencil and paper, desperate to invent double glazing before the Shieldhall's next trip.

"The pistons of the Shieldhall's engine are completely mesmeric. I'm surprised people don't fall into them and get crushed to death."

profits, merchant shipowners were were no more romantic in the nineteenth century than they are today. The enigma becomes all the more intriguing when you realize that, though the first steam-assisted crossing of the Atlantic was made by the USS *Savannah* as early as 1819, it wasn't until 1838 that the first crossing of the Atlantic was made by a ship travelling entirely under her own steam.

The answer becomes clear if we examine the case of the *Savannah* more closely. Crossing the Atlantic from New York to Liverpool, the ship had used up all her coal before she reached Ireland. It wasn't that

STEAM VS SAIL
IN BRITISH NINETEENTH-CENTURY MERCHANT SHIPPING

Anybody who ever hankered after the days of Empire will have their cockles warmed by the fact that from 1850 right up until the turn of the century, the British share of world shipping tonnage never fell below the 40 per cent mark.

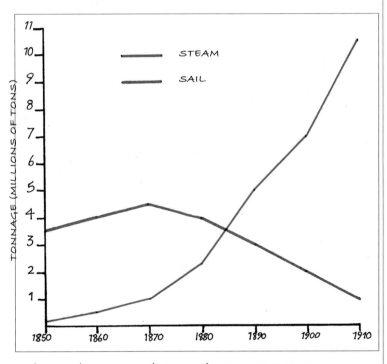

● In 1865 the first double expansion engine was put into service. By 1870 the tonnage of sailing ships had stopped growing while steam tonnage had more than doubled on the decade.

● In 1874 the <u>Propontis</u>, the first ship equipped with a triple expansion engine, was launched. By 1880 steam tonnage had shrunk on the decade while steam had once again more than doubled.

"Marvel, why don't you, at the wonderful hot thing between my legs. . . This is basically a little steam loco engine in a boat: it has one cylinder and steam is exhausted up the chimney. It's fine for footering around on the loch in, but to go any distance at all in this boat there'd be just about room for the engine, moi, and the coal and water I'd need to get me there. So if Auntie Bettie fancied a wee scoot along Loch Lomond the answer would have to be no.

Until the double expansion engine was successfully installed in a ship, steam posed no major threat to sail. A decade later, when safe boiler pressures had increased even further and the triple expansion engine was introduced, it was all over bar the shouting; steam navigation had arrived."

her captain had underestimated the quantity needed to power a steamship over such a distance – apart from food and water for the crew, she wasn't carrying anything *except* coal. The steam engine had come a long way since Newcomen in terms of fuel efficiency. Watt's engine was four times more efficient than Newcomen's, and Trevithick's best were three times as good as Watt's. But they still weren't good enough for long-distance passage-making. On the railways, engine efficiency was a simple matter of economics. A steam train could always stop at a coaling station to replenish its bunkers: steamships had to carry all they needed.

Water wasn't such a problem. Steam locos needed to replenish their boilers at regular intervals because they used the forced draught system to encourage the fire to spread forwards along the length of the boiler. Exhaust steam goes straight up the chimney and into the atmosphere, so a constant source of fresh water is required. But forced draught is only necessary because locomotive boilers have to be long and thin, and move in the opposite direction to the way you want the fire to move. Steam boats don't need forced draught, so they could easily employ Watt's separate condenser, improving the power of their engines by one atmosphere's pressure at the same time as saving the condensed water and injecting it, still hot, back into the boiler. But coal was another thing altogether.

The problem could be partially tackled by storing coal at foreign ports along the routes plied by the ships. This wouldn't help you cross oceans, but it would enable steam ships to be put to service on a lot

It's not often in the world of engineering that a picture so graphically tells a story. The high pressure is at 75 psi; the medium pressure sits at around 18psi and the low pressure sits at about 8 psi below atmospheric pressure. That's because every last drop of steam is being sucked out of the engine, sent upstairs, condensed into water and piped straight back into the boiler.

of other long-distance routes. And to some extent this was done. But there was a catch. The only way to get coal to foreign ports – many of which were thousands and thousands of miles from the nearest coal mine – was by ship. Shipping the coal over great distances increased its price considerably. This meant that for a long time steam was used only in cases where mail companies were willing to pay a premium in order to ensure their cargoes of letters got to their destinations quickly and predictably. That mail companies should have been the ones to foot the bill makes sense: it wasn't until steam navigation took off that shipping companies were able to provide anything resembling a timetabled service. Sailing ships, then as now, waited upon a favourable combination of wind and tide. In effect, the early steamships were subsidised by mail companies.

The innovation that finally enabled steamships first to vie with sail for precedence, and then to overtake it, was the compounding of cylinders. The compound steam engine was patented by Jonathan Hornblower as early as 1781. But Hornblower's engine used the expansive power of steam. Boulton & Watt sued him and won, and Hornblower discontinued his experiments.

Compounding involves putting more than one piston and cylinder together to form a unit. This doesn't sound that clever. After all, putting a lot of cylinders in a row instead of making a single cylinder bigger and bigger is a practice totally familiar to us from the internal combustion engine. One of the big reasons for using lots of small cylinders is the relative smoothness of the power output and reduction of vibration that this approach gives. But merchant ship owners weren't worried about vibration, they were worried about coal efficiency, and that's what compounding cylinders in a steam engine is about. The cylinders in a car are wired in parallel, each with its own individual access to the carburettor. But because steam can be re-used at lower pressures, steam engines offer a different

'wiring' possibility: the cylinders in a steam engine can be wired in series, the exhaust steam from each cylinder passing to the steam inlet of the next cylinder.

To understand why the compounding of cylinders was the innovation that brought steam power to the pitch of efficiency that allowed it to relegate sail to the position of romantic also-ran, you have to go back to first principles. Pascal was the first person to establish that as you go nearer to the surface of the sea of air we live in, the pressure of the air drops. He established this by sending his brother-in-law up a mountain with a column of mercury supported

"What I really love about ideas like this is that when you see them, you imagine that left on a desert island with twenty thousand pages of blank paper and a pencil you would come up with it yourself. And that's the sort of illusion that keeps us interested in machinery – because of course, it wouldn't happen really."

"It's funny looking at these and realizing I used to go to school with trains like this and yet now it looks like something out of a completely different era. It looks like a dinosaur really. I mean a very attractive dinosaur, of course . . . The kind of dinosaur you'd like to take to dinner and buy an expensive frock, and hope that someday, after a week or two, it would gaze at you in that wonderful, 'please – I love you – take me' way. Or am I thinking of something else? No, it's a train. It's definitely a train . . . Well, a locomotive. You must never call it a train. Guys in anoraks will beat you to death on the spot."

over a vacuum. And everyone knows that the temperature falls as you go up into the mountains. Jacques Charles established the relationship between the two phenomena: as the pressure of a gas increases or decreases, its temperature changes in direct proportion.

From the point of view of steam-engine design this fact of nature has a very undesireable side effect. If you put steam into a cylinder at, say, 150 psi, and go on to expand it to a vacuum, you are reducing the pressure of the steam – and therefore its temperature – a great deal. What this means for the steam engine is that on each stroke of the piston steam goes into the cylinder hot but ends up cold. So the cylinder cools, and when the next injection of steam comes much of the heat of that steam is used up in warming the walls of the cylinder back up. Using the whole of the energy of high-pressure steam in one big cylinder is an awful waste of energy.

The problem is similar to the fault Watt diagnosed in the Newcomen engine: cooling the cylinder to achieve condensation was a waste of heat, too. Watt's solution was to separate the cylinder and condenser and keep each at its optimum temperature. Compounding cylinders is another application of the same broad principle: by expanding the steam so far in one cylinder, then passing it to a second, larger cylinder for further expansion, it is possible to limit the temperature change inside each cylinder. Each cylinder can be kept at its optimum working temperature and very little steam energy is lost.

Brunel's *Great Britain*.

By 1843 Isambard Kingdom Brunel had brought the mechanical merchant ship to the point at which it was in all essentials a blueprint for the 'modern' steamship. Brunel was the last great engineer to have a finger in every engineering pie and his wide-ranging knowledge is reflected in his achievements in shipbuilding. Brunel had worked out that the resistance to the movement of a ship through the water did not increase in proportion to its tonnage. As ships get larger their carrying capacity increases more quickly than the fuel that is needed to propel them, and so larger ships have more space for freight. In other words, big ships are highly economical. He built a series of ships to prove his point. The *Great Britain* was the first with an all-iron hull. She weighed 3443 tons and was 322 feet long, and her iron hull was strong and inflexible enough to allow really powerful engines to be used. The design also incorporated the screw propeller, a far more efficient and reliable

"If there's too little water, it's more of a bomb than a boiler. . ."

system of propulsion than the old paddle wheels. This great vessel was in all essentials a thoroughly modern merchant ship. She lacked only one thing: a really efficient power-unit to turn the propellers.

In 1865 that that power unit arrived. Alfred Holt, a Liverpudlian civil engineer turned shipowner, built three iron ships equipped with high-pressure compound steam engines sufficiently compact, powerful and efficient that the ships could carry a cargo of 3000 tons – twice as much as the largest sailing ships of the day – at ten knots for over 8000 miles without having to refuel. The clever money immediately transferred to steam.

Holt's compound double engines were using steam at around 60 psi. By 1885 the compound double was superceded by the triple expansion engine. Triples used steam at pressures up to 125 psi, expanding it in three separate cylinders before condensing it and consigning it back to the boiler. By the end of Queen Victoria's reign

steamers were using steam at 200 psi and exploiting its energy so efficiently that, fully laden, the energy they used to carry a ton of cargo over one mile was no more than that generated by burning one sheet of good writing paper. Steamships were able to carry more cargo, and carry it faster, cheaper and more reliably than the best sailing ships. The battle between steam and sail was cancelled due to lack of interest.

So by the end of the nineteenth century steam-powered ships were plying all the world's great trading routes. Even the biggest, heaviest industrial products – such as steam locomotives – could be shipped efficiently and cheaply to faraway places. And so the great expansion

"Now, I'm not a guy who lives in the past – indeed I embrace the present with some enthusiasm. I love my electric toothbrush . . . I like my personal computer . . . and where would I be without Nurse McKenzie's patent electronic virility accelerator? But having said that, is there a heart so base that it doesn't flutter – nay, thunder, at the mere sight, sound and smell of these beauties – and break a little in the certain knowledge that we will never see their like again?"

of the railways that had been taking place in Europe and America throughout the nineteeth century could continue across continents where no locomotives were being manufactured. Steam had conquered the world.

CHAPTER THREE

TWO-STROKE

There are people, some of whom are knowledgeable and passionate about engineering, who go through life burdened with a barely concealed hatred for the two-stroke engine. And on the other hand there are people, no less well informed or committed to the subject, who find in the two-stroke more to admire and fascinate them than in any other type of engine. In much the same way, opinion is divided about rats and other vermin: while some people are filled with admiration at their ability to adapt and find a niche for themselves in any circumstances, others find that precisely the same phenomenon fills them with distaste. I'm not one to sit on the fence on these matters, but for the purpose of discussion, I am willing to suspend my judgement. And if something of my own opinion should slip out, please don't think it's because I'm trying to sway you.

"You can pick out a two-stroke engine a mile off by the noise it makes – something like a swarm of hornets with the hiccups, only less homely sounding. And then there's the dirty cloud of oily blue exhaust smoke. The two-stroke engine didn't so much turn the world as totally infest it. And personally, I hate the buggers."

But though I may be opinionated, prejudiced I am not. Because when, in an atmosphere of calm – when, for instance, I am sitting behind the wheel of a quietly burbling four-stroke V8 – I clear from my mind the two-stroke's irritating whine and reflect on what's actually going on inside, I have to admit that it is a brilliant piece of engineering. The two-stroke engine has a concision and an organic simplicity that merits comparison with the greatest products of the human imagination. It achieves everything that needs to be achieved by an internal combustion engine with so few moving parts that if it hadn't already been invented you could be forgiven for thinking it wasn't possible.

So what is it about the two-stroke that makes it simultaneously so clever and so irritating? And why do we have two different kinds of internal combustion petrol engine at all? Come with me, why don't you, on a journey of discovery . . .

"The thing about vermin is when you chase them away they always make a home somewhere else. I have a sneaking suspicion that the engines that used to power two-stroke motorbikes have simply decamped and made new homes for themselves in jet-skis, where they are enthusiastically lowering the tone at lakes and beaches all around the world. Other habitual lairs include microlight aircraft, radio-controlled planes, racing bikes, hovercraft, hedge-strimmers, snowmobiles, motorboats, scooters, go-karts, hydroplanes - anywhere where mechanical simplicity and a high power-to-weight ratio are more important than good taste, in fact."

To begin at the beginning, the simplest and most ancient internal combustion engine is the cannon. The barrel of the cannon acts like a cylinder and the cannonball like a piston. To get from a cannon to a self-sustaining internal combustion engine producing smooth, continuous rotary motion you have to do quite a number of modifications. Aside from the experimental gunpowder-fired atmospheric engine developed by the Dutchman Christiaan Huygens in the seventeenth century (see *Steam*), nothing much happened in the field until the middle of the nineteenth century. The steam engine was nearing the limits of of its efficiency at this time and engineers were becoming aware its intrinsic disadvantages: coal, fresh water and a heavy boiler means they're always going to be rather weighty for their power output; the need to get up a head of steam before doing any work makes them laborious to operate; and even in their most perfect manifestation, they still gobble up an awful lot of coal.

In 1860 the Frenchman Jean Joseph Etienne Lenoir patented an internal combustion engine that ran on a mixture of coal-gas and air. Lenoir's gas engine was the first half-way decent internal combustion engine. It was quiet and three times as fuel-efficient as the most efficient steam engine. Like the steam engines of the day Lenoir's engine was double-acting. Combustion happened successively on either side of the piston, so every stroke of the piston was a power stroke. Lenoir's engine created a big stir. Several hundred of them were built and the inventor himself succeeded in powering a road car with one. But it wasn't very powerful and offered no real competition to the steam engine.

Other engineers continued to develop Lenoir's ideas. The German Nicholaus Otto developed a particularly good version. It was more efficient than Lenoir's original and very commercially successful. But as Otto continued to develop his Lenoir-type gas engine he became frustrated that it could not be made to yield more power. Three horsepower seemed to be the practical limit.

Otto discovered that the only way of improving on the low power output of internal combustion engines was to compress the gas/air mixture before igniting it. He had actually discovered this principle during his earliest experiments with internal combustion engines, long before he had developed his Lenoir-type gas engine. But at that time the violence of the explosion produced when gas and air are compressed before ignition had seemed like a problem rather than a bonus, and the difficulty of containing such unlooked-for power deterred him from developing the compression engine further. But when his Lenoir-type gas engine wouldn't yield any more power, Otto remembered his early experiments with compression and realised that this extra release of energy could be the solution to his problems.

So, while other nineteenth-century engineers were still thinking of the internal combustion engine in terms of steam engine design, and were intent on mimicking the best steam engines of the day by making each stroke of the piston a power-stroke, Otto came up with the idea of stretching the whole process over four strokes of the piston - one of which would compress the gas/air mixture before it was ignited. This wild deviation from current thinking paid off handsomely. The extra

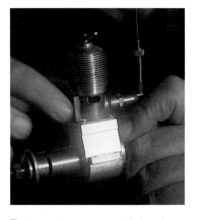

The two-stroke engine really is quite a simple thing . . . When the piston goes up to fire, the vacuum inside the crankcase is used to draw fuel and air mixture through the carburettor, and when it comes back down, it pressurizes the crankcase and pushes the fuel up and on to the top of the piston, so the piston comes up and fires again . . . and it just goes on and on and on.

THE
FOUR-STROKE
CYCLE

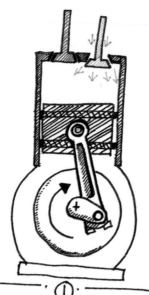

① INDUCTION

INLET VALVE OPEN. PISTON PULLED DOWN BY CRANK DRAWS AIR/FUEL INTO CYLINDER.

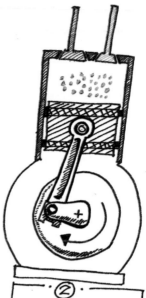

② COMPRESSION

BOTH VALVES CLOSED, PISTON COMPRESSES MIXTURE, IT HEATS UP.

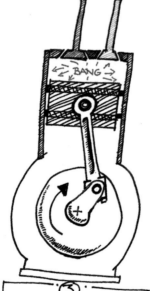

BANG

③ IGNITION

BOTH VALVES STILL CLOSED, SPARK PLUG IGNITES MIX, EXPLOSION DRIVES PISTON DOWN.

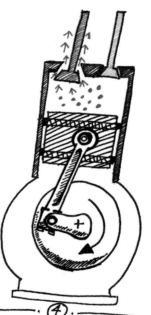

④ EXHAUST

EXHAUST VALVE OPENS, PISTON SLIDES UP AND DRIVES BURNT GAS OUT.

power yielded by high-pressure combustion was more than enough to compensate for the extra work done by the engine.

Otto's introduction of the compression stroke was responsible for making the internal combustion reciprocating engine what it is today: the prime mover of the twentieth century. He built his first four-stroke internal combustion engine in 1876 and in 1877 produced his first petrol-burning four-stroke. With the use of petrol rather than gas – which has to be piped, and so limited the engine to static applications – the four-stroke cycle had really arrived. Demand for his four-stroke was so overwhelming that production of his Lenoir-type gas engine stopped immediately. Eight thousand four-strokes were produced in the next few years.

Though it is mechanically the simpler of the two engines, the two-stroke was invented after the four-stroke. And despite its greater mechanical complexity, the four-stroke is really much the more straightforward of the two designs. Each of the the piston's four strokes accomplishes a separate function efficiently and in proper sequence. On the combustion stroke, rapidly expanding gases produced by igniting the fuel/air mixture drive the piston down the cylinder. This is the power stroke. The compression stroke is there to take advantage of Otto's great discovery, compressing the fuel/air mixture before it is ignited. The induction stroke is dedicated to drawing fresh fuel/air mixture into the cylinder, and the last of the four, the exhaust stroke, expels the burnt gases that are left after combustion has taken place. The order of the strokes is: induction of fresh fuel/air mixture; compression of that mixture; combustion of the mixture and expansion of the resulting gases; exhaust of the burnt gases. This cycle is repeated over and over for as long as the engine is running.

Dedicating one stroke of the piston to each of these four functions is eminently sensible and practical. But looked at another way, sensible as it is, the Otto cycle only gives one power stroke in four. That means that three quarters of the time the pistons are not driving the crankshaft but being driven by it. If you could find a way of having an engine perform induction, compression, combustion and exhaust all in two strokes of the piston then your engine would produce twice the number

The two-stroke petrol engine was developed after the four-stroke had already been perfected – very much with marine applications in mind. When you choose an engine to put in a boat, power-to-weight ratio is terrifically important. In small vessels a lighter engine means a shallower draught and so more flexibility in shallow waters. With bigger ships, you can cash in on any space and weight saved in the engine room by carrying more cargo. That's why the two-stroke is still the marine engine of choice, from tiny outboards to the massive cathedral diesels that power supertankers.

Two-Stroke

①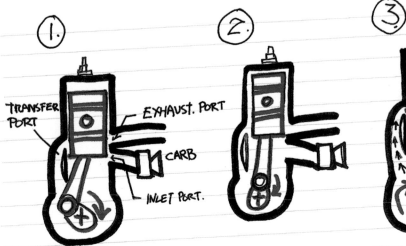

TRANSFER PORT

EXHAUST. PORT

CARB

INLET PORT.

②

③

1. 'As the piston comes up a partial vacuum is created in the crankcase. When the inlet port is uncovered fuel/air mixture rushes into the crankcase.'

2. 'As the piston reaches the top the spark plug sparks and the explosion drives the piston down.'

3. 'The piston reveals the exhaust port, allowing the exhaust gasses to escape. A moment later the transfer port is revealed, allowing the pressurised mixture to transfer itself from case to cylinder. The genius and fundamental problem of the 2-stroke is that fresh fuel is arriving while the exhaust gasses are still putting their coats on. A choice of symptoms: half the fuel blows out of the exhaust port, or half the exhaust remains to pollute the air/fuel mix.'

④

4. 'Piston on its way north again, cutting off exhaust and transfer ports. Above the piston fuel is being compressed; below a vacuum is being created. Some 2-strokes do this 22,000 times a minute. Scary, or what?'

of power-strokes of the Otto cycle at any given engine speed. Your engine would then be half the size, half the cost and half the weight of a four-stroke and still be able to do the same amount of work. Not an unattractive proposition.

But could you really eliminate half of the Otto cycle and still have a working engine? You obviously wouldn't want to get rid of the power stroke. And the compression stroke, though not absolutely necessary, is extraordinarily good value, releasing far more energy from the fuel than the mechanical energy it absorbs. The only possible slack lies in the two scavenging strokes. And appropriately enough, it is in their method of scavenging that the verminous two-stroke and the stately four-stroke differ.

Dispensing with the exhaust stroke is quite straightforward. The fastidious four-stroke opens a valve and pumps the exhaust gases out of the cylinder with one up-stroke of the piston. But those hot exhaust gases are already under much greater pressure than the atmosphere. Why bother to pump them at all? Given half the chance they'll escape from the cylinder of their own accord. That's just what happens in a two-stroke. Towards the bottom of the power stroke an exhaust port – basically just a hole in the side of the cylinder – is revealed by the piston. The high-pressure gases simply flow out of the cylinder and away down the exhaust pipe.

That wasn't so hard. And if that was all there was to it, the internal workings of the two-stroke engine wouldn't be much to write home about. But what about those engineers who become obsessed by the two-stroke, and devote their working lives to mastering its subtleties? In order to get an inkling of what fascinates them, you have to look at the way the two-stroke gets the fuel/air mixture into the cylinder. Because, loath though I am to say it, the two-stroke intake system is an inspired piece of work.

In a four-stroke, as we've seen, the fuel/air mix is pumped into the cylinder by the piston. The piston moves down the cylinder, creating a partial vacuum and so sucking in fuel/air mixture from the carburettor. But there's no place in the two-stroke's action-packed cycle for the piston to perform this function. And unlike the pressurized

---

ENGINES THAT DIDN'T TURN THE WORLD
THE SIX-STROKE CYCLE

Four-stroke not fastidious enough for you? See if you can get your hands on a venerable old British six-stroke engine. The six-stroke only produces one power stroke for every six strokes of the piston - that's three whole rotations of the crankshaft. The two extra strokes are used to draw in clean air from the atmosphere and flush it back out again, ridding the cylinder of every last trace of exhaust gases before the fresh fuel/air mixture is drawn in from the carburettor. If you like novelties and don't care a hoot about the weight, size, cost or complexity of your engine, it's great. Its mechanical efficiency is only five or six percent less than that of a four-stroke - that's the cost of having the piston do that extra work - but its overall fuel efficiency is almost the same as the four-stroke, so much more powerful is combustion when all traces of exhaust gas have been removed.

PISTON PORTS

There's more than one way of arranging that a four-stroke engine's fuel inlet and exhaust valves should open and close at the appropriate moment, but all the systems are fairly complex; cost money to manufacture; need lubricating; take power from the engine; and limit the engine's maximum speed, and therefore its power output. By using piston ports the two-stroke overcomes these disadvantages. Piston ports are opened and closed at precisely the right point in the cycle every single time, for the very good reason that they work by being covered and uncovered by the piston itself.

exhaust gases, the fuel/air mixture is at atmospheric pressure, so it's not going to go anywhere of its own accord. What the two-stroke needs is a separate pumping mechanism. Now, it would be possible to rig up a pump for just this purpose. You could use four-stroke type poppet valves to let fuel mixture into the cylinder at exactly the right time and drive the whole caboodle off the crankshaft. This is just what the very first functioning two-stroke engine did. It was built in 1879 to a design by the Glaswegian engineer Sir Dugald Clerk. But a separate air-pump is hardly an elegant engineering solution. It involves an increase in overall mechanical complexity, with a consequent increase in manufacturing cost, engine friction and maintenance requirements. Not much to get excited about at all.

But there's a better way. The piston and cylinder can be used as a pump in the four-stroke cycle because the cylinder is pressurized. This just means it is sealed to the atmosphere, so when the piston moves down the cylinder a vacuum is formed, and when it moves upwards the gas in the cylinder is compressed. Now in a four-stroke engine the space at the bottom of the cylinder – the crankcase – is used for nothing more exciting than holding a reservoir of oil to lubricate the engine. As long as you can find a different way of lubricating the engine, there's no reason why the space below the piston shouldn't be pressurized as well as the space above. In this way the piston can do work at both ends, and the bottom end of the piston can take over pumping duties from the top, leaving the top end free to pack in those extra combustion strokes.

So that's your two scavenging strokes abolished and, in principle, expulsion of burnt gases and induction of fresh fuel are both being organised in good time for the next compression stroke. The pay-off is that the engine delivers power once in every two strokes of the piston, so at a given engine speed the two-stroke piston delivers twice as many power-impulses to the crankshaft as the four-stroke. And there's a bonus prize too: the inlet and exhaust valves and all their driving machinery are now completely unnecessary, bringing a further saving in weight, manufacturing cost and engine drag. So for any given power-requirement you can reduce the size and weight of your engine by half.

Even a man of confirmed tastes, from which the two-stroke engine is notably absent, would have to admit that that kind of power-to-weight prowess can't be a bad thing. It's easy to see how the verminous two-stroke proliferates in applications where size and weight are especially important. You might even be forgiven for wondering why the two-stroke hasn't been more widely successful. How come it hasn't found a home in our cars and lorries, for instance?

Well, the truth is the two-stroke engine has problems that go far beyond the aesthetic disadvantage of sounding like an over-stressed hairdryer. The pressurized crankcase is indeed one of engineering's great short-cuts. But the problem with short-cuts is they don't always bring you out exactly where you want to be. When you separate intake and exhaust and look at the way the two-stroke handles them then everything seems fine. But put the two parts of the scavenging process back together and you are immediately faced with a thorny problem: in the two-stroke engine, intake and exhaust take place simultaneously. The exhaust port is uncovered before the inlet port, so that when the fuel/air intake is opened the pressure in the cylinder is low enough to allow fresh fuel/air mixture to enter. All well and good. But then it remains open for the entire time the transfer port is letting in fresh fuel/air mix, and – since it is positioned above the transfer port – for a short while afterwards as well. Inevitably, a proportion of the fresh fuel/air mixture is pumped straight through the combustion chamber, out of the exhaust port and into the atmosphere. This is why two-stroke engines are currently being persecuted by governments all around the globe. Two-stroke motorbikes were decimated by USA emissions controls, two-stroke outboards are currently being banned on inland lakes across Europe and America. Even in developing countries like India where problems of pollution are low on the political agenda, two-stroke powered rickshaws are coming under severe scrutiny.

This loss of fuel and consequent pollution of the environment can be limited to some degree by judicious placement of the transfer and exhaust ports, and by using a shaped piston in order to direct the incoming fuel/air mixture away from the exhaust port. But at the end of the day you have to make a choice between two unacceptable options. On the one hand you can reduce the amount of fuel/air mixture coming in through the transfer port so that it doesn't have time to escape

"The best thing about the two-stroke is the idea. As an abstract engineering feat it's sans pareil. It's one of those things that shouldn't work . . . It's like the bee, it shouldn't fly. And it gives cheap pleasure to lots of people, and you can't knock that."

though the exhaust port before it is closed off by the piston. This sounds fine in principle but in practice results in a fuel/air mixture so diluted by exhaust gases that the engine's power is radically impaired. And on the other hand you can allow enough fuel/air mixture into the cylinder to help drive out the remaining exhaust gases and give a rich mixture and full combustion - and pump unburnt fuel out into the atmosphere in the process.

The second reason for the two-stroke's reputation as a polluter also has to do with the use of the crankcase as a gas pump. If you were to continue to use the crankcase in its traditional role as oil reservoir then rather a large amount of oil would find its way into the combustion chamber, causing all sorts of problems. Instead, the two-stroke engine is lubricated by adding a small amount of oil to the petrol. This oil vaporizes in the carburettor along with the petrol and, as the mixture

TWO-STROKE SAAB

It's because of the two-stroke's system of lubrication that the two-stroke Saabs that dominated the rally scene in the early sixties had a freewheel in the gearbox. Drive a two-stroke car down a long hill in the usual fashion, with the throttle closed and the car in gear, and you're going to get into problems. Your crankshaft will be rotating and your pistons pounding up and down at full speed, but because the throttle's closed there'll only be the faintest breath of fuel/oil/air mixture circulating round the engine. In the absence of oil, the friction between metal surfaces increases drastically. The pistons and cylinders get very hot indeed and after not very long at all you may find that your cylinder block and pistons, once separate individuals, have suddenly been transformed into mere aspects of a greater unity - that is, a useless lump of scrap metal. This process is called engine seizure and the freewheel gear was invented to stop it. When it is engaged the wheels no longer drive the engine. The engine runs at a tick-over and a small amount of fuel/air/oil mixture provides adequate lubrication.

JOSEPH DAY

No one man is popularly associated with the invention of the two-stroke engine. Well, would you have owned up to it? Exactly. But I've tracked him down. His name is Joseph Day. He's the man who built his first successful piston-ported, crankcase scavenging two-stroke engine, way back in 1891. It's true that Sir Dugald Clerk had already patented the two-stroke cycle in 1878 and built the first full-size working engine in 1879, and that in Britain at least the two-stroke cycle was known as the Clerk cycle for many years. But Clerk's engine ran on coal gas and used a separate air-pump to charge the cylinder. Joseph Day was the first person in Europe to build the kind of two-stroke that can spoil a pleasant Sunday afternoon. He combined the crankcase scavenge system that had been developed by John Fielding in 1881 with the simple piston-port system of expelling the exhaust gas, so producing the first two-stroke engine that ran without valve machinery or a separate scavenge pump. By 1906 petrol-fuelled two-strokes were being sold by the Day Motor Co. Ltd, Putney. They had power outputs ranging between 2.5 to 30 hp and were sold mainly for marine use.

circulates around the crankcase and combustion chamber the oily petrol vapour provides effective lubrication for the engine's major moving parts. It's better than burning the vast quantity of oil that would find its way into the combustion chamber via the transfer duct, but it's still not good. It means that on top of all the usual muck the exhaust gases contain rather a lot of burnt oil. That's why the exhaust smoke of two-stroke engines has that characteristic blue hue. It also means that the fuel/air mix that escapes straight into the atmosphere is enriched with a quantity of unburnt oil. With the introduction of synthetic oils, the amount of oil in the petrol has gone down over the years - from a ten per cent solution in early engines to as little as a two per cent solution - but however little you put in to start with, burning oil with your petrol is a pretty poor way to lubricate an engine.

So, as I was saying before I came over all open-minded, the two-stroke engine is a nasty piece of work. Atmospheric pollution is a problem that isn't going to go away and the two-stroke's striking advantages as a light-weight power unit are more than outweighed by its dirty habits. And I know I've already mentioned this, but the noise it makes is just awful. But to be fair, there is something that can be done about the problem of that gaping exhaust port. It's the two-stroke engine's most arcane secret: the expansion chamber exhaust pipe. The expansion chamber really is a marvel of imaginative engineering. If you were even moderately impressed by the pressurized crankcase scavenging system, you're going to love it. And there's a good story behind it too.

If, like me, you associate the two-stroke primarily with the bright, brash and aggressively free-market Japanese motorbikes of the seventies and eighties, you might be surprised to hear that the two-stroke engine was once an avowed socialist. If there was ever an engine of the people it was the two-stroke. The reasons for this are quite straightforward. There's no simpler engine to build or maintain than the two-stroke, since crankcase scavenging and piston ports make most of the four-stroke's moving parts unnecessary. Take away the camshaft, pushrods, rockers, tappets, valves, oil pump, oil filter and a few more bits and pieces, and the internal combustion engine starts to take on a whole new aspect. It becomes easier and cheaper to manufacture, easier and cheaper to maintain, it can be worked on without specialised knowledge or special tools, it requires servicing far less often than its upmarket cousin and, because it is smaller for its power output, it uses up less valuable metal too. A pretty convincing list of advantages if you're interested in bringing power to the people in a nation which isn't economically advantaged.

And so after the war the two-stroke became the engine of choice in East Germany – engine builders to the Soviet Bloc. East Germany was an industrially developed but newly poor country, ambitious for that peculiarly socialist kind of upward mobility in which you don't move to a better area but instead wait for the government to arrange for one to be delivered. At the IFA engineering works at Zschopau, just to the east of the Czech border, the forty or so car and bike manufacturers that had existed in Germany before the war were brought together under one state-run roof. Cheap, functional cars and motorbikes were designed and built there, including the now legendary Trabant and the upmarket Wartburg – the wheels of choice for state officials all over eastern Europe. It was at Zschopau that the MZ motorbike was made.

TRABANT

Since the Berlin Wall came down the two-stroke Trabant has been the butt of so many jokes that a word or two ought to be said in its defence. It was designed to make personal transportation available to as many people as possible in a country that was very wealthy by Eastern bloc standards but outstandingly poor when compared to its Western neighbours. Yes, the bodywork was made out of cardboard – but it didn't rust, had a two-year service interval, was exceedingly reliable, could be taken apart with a handful of spanners and provided practical, inexpensive transportation for people all over Eastern Europe.

"Don't knock the Trabby. It's reliable, efficient, and costs about half as much as your car to run. And the entire car can be dismantled with four spanners."

At the other end of the scale were the large luxury Wartburgs – the car favoured by Eastern bloc politicians. But just like the Trabant, because they're powered by a two-stroke engine it is quite possible for one man to lift the entire engine out from under the bonnet and fix it by the side of the road with a minimal toolkit, without the help of an engine hoist.

"The great thing about the Trabant is that it's full of all the individualism you saw in the Chevy, but they didn't have the money. All these great little paint jobs and funny wee lights and guys with leopardskin seats. I thought it was a tribute to the human spirit. They said, 'Okay, if this is what you're going to give us, we're going to make it fun.'"

"I wonder if people steal these engines. They're so easy to get out . . ."

Like Trabant and Wartburg, MZ has a reputation in the West for producing motorised rubbish. But for many years MZ were at the cutting edge of two-stroke technology, and for a while – until the Japanese stole their thunder – they were the dominant force in grand prix bike racing. The reason for their success? They took the two-stroke seriously when, because of the innate inefficiencies discussed above, it was dismissed across the rest of Europe as an engineering dead end. The engineer responsible for turning the two-stroke from a petrol-spewing good-idea-in-principle into the terrifyingly powerful racing engine it is today was one Walter Kaaden.

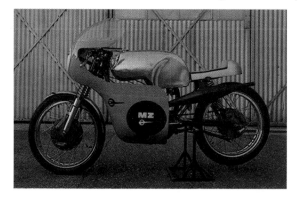

Kaaden was born in 1919. One of his less endearing claims to fame is that he worked for the Nazi government developing the V1 and V2 flying bombs. When Germany was defeated Kaaden was interned by the Americans before returning to East Germany. He was working on his own, tuning up motorcycles, when he was talent-spotted by the IFA factory. By 1953 he was running their racing department. By 1955 he had developed his version of the two-stroke engine to the point where it was performing almost as well as the best of the four-strokes. Kaaden made many important modifications to the two-stroke but by far the most important was his work on the two-stroke exhaust. As you will remember, the main problem with the two-stroke is that the exhaust port remains open while the fuel/air mix is entering the cylinder. So either you pump fresh fuel straight out of the exhaust, or you draw less fuel/air mix into the cylinder – and satisfy yourself with a hopelessly weak mixture and low power. The compression ratio is also adversely affected, since instead of starting right at the bottom of the cylinder, the compression stroke doesn't start until the exhaust port is closed. Kaaden's big idea for overcoming these problems was to produce a very special kind of exhaust valve. And when I say very special, I'm not exaggerating: because the exhaust valve Kaaden invented is made out of exhaust gas.

The expansion chamber exhaust pipe is the device that makes this possible. When the exhaust port of a two-stroke engine is uncovered by the piston, a shock wave of high-pressure exhaust gas is emitted

"Walter Kaaden took the two-stroke engine and turned it from a lawnmower into the fastest motorbike in the world."

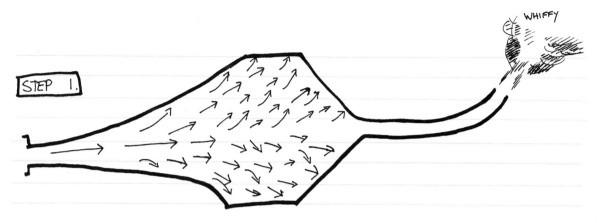

WHIFFY

STEP 1.

Step 1: High pressure shock-wave scoots down pipe, reaches expansion section, expands, drops in pressure, sucking more exhaust gas out behind it.

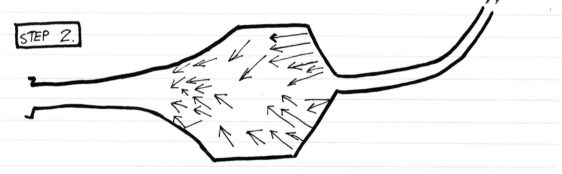

STEP 2.

Step 2: Shock-wave hits constriction in pipe and is reflected back towards the port, where it temporarily blocks the exit for fresh fuel/air mixture until the piston arrives and does the job properly by covering the port.

This all happens in microseconds. Applied skilfully it's a trick that can double the power output of a two-stroke engine for no more than the cost of an exhaust pipe shaped like a market vegetable. The trick lies in which market vegetable you choose to model it on.

"You've got to hand it to the Germans when it comes to efficiency allied to inventiveness. The history of engineering is littered with them."

from the port into the pipe. The expansion chamber pipe works by manipulating this shockwave and putting it to work. In a straight exhaust pipe the shockwaves follow one another along the length of the exhaust pipe and out into the atmosphere in a regular and altogether uninteresting fashion. But if the diameter of the pipe is varied, the waves can be reflected back and forth. Increasing the diameter of the pipe causes low-pressure reflections; decreasing the diameter causes high-pressure reflections. By carefully arranging first an expansion and then a contraction of the diameter of the exhaust pipe, Kaaden discovered that he could use a low-pressure reflection to help suck the exhaust gas from the cylinder, then a high-pressure reflection to plug the exhaust port until it was closed by the piston, so reducing the loss of fresh fuel/air mixture. Since less gas is lost compression is also improved. This arrangement gives a massive boost

WALTER KAADEN

achieving with his engines; the only way he could continue to exploit the advantages of his advances in two-stroke technology was by bartering with a British grand prix race team – a shipment of his own high-quality drum brakes in return for a load of British-made Norton racing forks.

Kaaden's great discovery allowed MZ to win on the grand prix circuit even though their bikes were nowhere near as well manufactured as their competitors. MZ didn't have enough foreign currency to buy high-quality components so Kaaden had to resort to barter: MZ's forks were too weak for the power output that Kaaden was

The 1961 Swedish Grand Prix signalled the beginning of the end for the German, American and British bike industries.

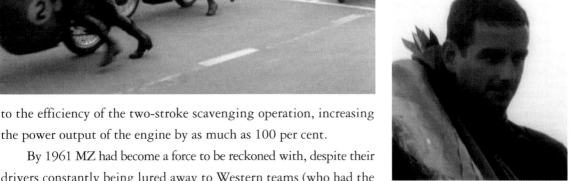

to the efficiency of the two-stroke scavenging operation, increasing the power output of the engine by as much as 100 per cent.

By 1961 MZ had become a force to be reckoned with, despite their drivers constantly being lured away to Western teams (who had the advantage of being able to pay them money). The power-to-weight ratio of Kaaden's race-tuned two-strokes was staggering. Their 125cc two-stroke could outperform 350cc four-stroke bikes; their 250cc two-stroke had a top speed of 155 m.p.h. and could match the best of the 500cc bikes. And Walter Kaaden was achieving this minor miracle on a budget of pennies. At the Swedish Grand Prix – the penultimate race of the season – it was looking very much as if MZ might win the 125cc

Ernst Degner, MZ's star rider, defected to Suzuki in 1961. His career with Suzuki was ended by a crash in which he sustained horrific burns. After a long and painful recovery Degner joined Kawasaki. He crashed again, this time suffering serious head injuries. Degner took the hint and retired from racing. He died years later in obscurity. It is said he committed suicide.

Alan Shepherd rode for MZ alongside Ernst Degner and still remembers the impact of Degner's defection on the team: 'Had Ernst Degner finished in the 125 race in the Swedish Grand Prix, Walter (Kaaden) would have got first place in the motorcycle road race World Championship.'

THE BSA BANTAM

The BSA Bantam – the English postman's favourite ride. Four hundred thousand of them were produced in total. The GPO bought a fleet of them for telegram delivery. A great British success story? Not quite: the BSA Bantam was the pre-World War Two DKW  RT125. DKW were one of the companies that came together under the IFA umbrella. The Germans gave the design to the British as part of war reparations. BSA put on a British flywheel magneto, a new carburettor, converted the nuts and bolts from metric to imperial and produced a mirror image reversal of the whole engine, which had the result of placing the gear lever and kickstart on the right-hand side, as it should be on a British bike. BSA kept the origin of their popular little runabout top secret. The Bantam had a 123cc engine, producing 4.5 bhp at 5000 r.p.m. It weighed 77kg and had a top speed of a staggering 53 m.p.h. It had three rather widely-spaced gears, and consequently had some difficulty in getting up hills.

championship for the first time. Their top rider, Ernst Degner, had already won three races that year, and the bikes were getting better and better. When the race kicked off Degner went straight into the lead. After only two laps he was 200 yards ahead of the next man. An amazing performance, marred only by the fact that half a lap later Degner drove up a slip-road, parked his bike and left the track with a friendly man from Suzuki, arranging to take enough trade secrets with him to enable Suzuki to copy – and soon improve upon – the main points of Kaaden's brilliant design. Within ten years the Japanese manufacturers came to dominate grand prix racing. As we all know, they also went on to wipe out the British and American motorcycle industries.

Now you might be forgiven for thinking that the two-stroke's troubles were over: masses of cheap, lightweight power, and fuel spilling into the exhaust a thing of the past, all thanks to Kaaden's expansion chamber pipe. Well, not quite. The expansion chamber exhaust

VESPA SCOOTER

The first Vespa scooter was built in 1946. It was partly motivated by the pressing need for cheap personal transportation in post-war Italy, partly by the need of aircraft manufacturer Piaggio to diversify. It was designed by Corradino d'Ascanio around a 98cc two-stroke aircraft starter motor. For convenience and comfort – and in order to make his motorcycle attractive to people who, despite all its

It was futuristic . . . It was beautiful . . . And everybody wanted one.

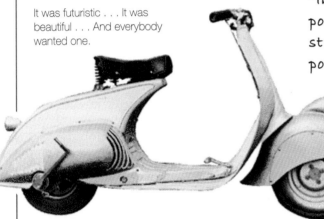

"In the economic grimness of the postwar Eastern bloc the two-stroke brought cheap, efficient power to the people. But in the West, the emphasis was on pleasure. It was in Italy that they finally came up with a use for the two-stroke that caught people's imagination. In fact, it sold by the millions."

advantages, stubbornly refused to be seen outside the house in leather gear – d'Ascanio designed the body as a pressed steel monocoque. The clutchless gears were operated by a twist-grip. In 1948 the engine was uprated to 125cc and in 1954 a 150cc model was introduced. Piaggio, who started out in Genoa in 1884 as a manufacturer of ship fittings, became the world's third largest manufacturer of motorbikes. D'Ascanio's scooter was named in honour of the dreadful buzzing sound its two-stroke engine makes: vespa is Italian for wasp.

"Meanwhile, in Britain, the fashion was for model-making . . .and the two-stroke was just perfect. Here was an engine whose mechanical parts were so simple they could be reproduced in miniature, and they'd still work."

"In the US, they like their toys a little bigger"

revolutionised the two-stroke's potential as a racing machine but it doesn't do as much as you might think to clean the engine up. Because, like an organ pipe, the expansion pipe is tuned to resonate at a particular frequency. This frequency corresponds to a particular engine speed. When the engine is running below that speed the high-pressure reflection reaches the exhaust port too soon and, instead of preventing the loss of fuel/air mix through the exhaust, actually inhibits the outward flow of exhaust gas instead. This results in a weak mixture and inefficient combustion.

When the engine is running above the resonant frequency, the high-pressure reflection gets there too late to do any good, plugging up the exhaust port when it has been physically closed already, and so doing nothing to prevent the escape of fuel into the atmosphere. The pipe works to a useful degree across a fairly large range of engine speeds – about 3000 r.p.m. – but when it isn't working we're back with the dirty old scavenger again . . .

SNOW-POWER

When tuned two-strokes are used to power vehicles their narrow peak power band gives them stunningly uneven acceleration and makes them thrilling to ride. The snowmobile is possibly the ideal application for the two-stroke engine. The light weight of the two-stroke makes the snowmobile more effective on soft snow and safer on thin ice. And in sub-zero temperatures, the two-stroke's method of lubrication makes perfect sense: oil left to stand in the four-stroke's sump solidifies at low temperatures and has to be warmed through before it can provide effective lubrication. And, finally, of course, out in the Arctic tundra there's far less chance of the two-stroke's high-pitched buzz annoying the neighbours. The snowmobile isn't just a toy, by the way; it has transformed the lives of people who live in the extreme North in just as spectacular a way as the motorcar has for people in warmer climes.

THE TWO-STROKE FIGHTS BACK

The Mercury Orbital V6 outboard motor manufactured by Mercury in the USA is just about the most advanced production two-stroke in the world. Crankcase scavenging is out; on-board computers with more calculating power than the average lap-top are in. The computer constantly monitors the temperature and chemical composition of exhaust emissions. This information is used to make decisions which are fed to the fuel-injection system. The injector can vary the timing and size of the fuel injection every single cycle. Even the direction of the injection, its shape and the size of the droplets of petrol have been meticulously calculated to improve emissions performance. And it works; it runs reliably and smoothly and is as clean as a four-stroke. The major car companies are all investigating the possibility of exploiting the new two-stroke technology. Oh dear. It looks like they might be making a comeback.

Throughout the eighties the two-stroke Yammy RD was the coolest thing a teenage boy could wrap around a lamppost. Good riddance to it. The two-stroke motorcycle has died a death in this country because of the United States' emission controls, which in some states are set at a level that even a well-tuned two-stroke cannot meet. As far as the motorcycle manufacturers are concerned, if the US don't want it, we can't have it in Britain either. What a shame . . .

"The chainsaw is the one justification for the two-stroke engine – because no matter how painful it may be to one's ecological sensitivities, we need wood".

. . . As I was saying, the two-stroke is a noisy, filthy little whipper-snapper of an engine, and the sooner some effective method of extermination is devised the better. Interesting idea, though.

CHAPTER FOUR

DIESEL

The story of Rudolf Diesel could almost be retold in the style of classical myth. There was once a brilliant young man who loved engines. One day he heard a description of an engine that was superior to all existing engines – an ideal engine that would turn every last drop of its fuel into useful work. He decided to devote his life to making this ideal a reality. He struggled so hard to realise his ideal that some of the gods decided that his super-human dedication deserved some reward. But other gods were annoyed at his insolence in presuming to dominate the forces of nature so completely. And so a compromise was reached, and – in that playfully malicious spirit that only the pre-Christian deities really understood – they turned the man into the engine of his dreams.

Rudolf Diesel

Of course, that isn't at all what happened. But it seems a bit like it because for most people Diesel is no longer the name of a man, it's the name of an engine and of the fuel it burns. For an engine as important and successful as the diesel engine to remain permanently associated with its inventor is unique in the history of engineering. Usually, the more successful an engine is the weaker its association with its inventor becomes. Why didn't this happen in Diesel's case? Possibly because, in so far as an engine can express a personality, the diesel engine appears to an uncanny degree to be an expression of the character of its inventor: rational, obdurate, working class – but nevertheless a big money-maker. This mythical account of the man and his engine is certainly incomplete: Diesel was truly passionate about engineering. He was an enthusiastic salesman, able to talk the hind legs off a donkey when the need arose. And he was a sensitive, cultured man. He was a great lover of art and music, sketched and played the piano, and had dreams of egalitarian, co-operative political reforms.

Rudolf Diesel's childhood was hard enough to have crushed the spirit out of many people. But he was a stubborn and determined child

"Rudolf was a driven man who throughout his life excelled in everything he did."

and never let the obstacles that lay in the way of his ambition deter him. And so he grew into a stubborn and determined man who wouldn't rest until he had made his dream of a theoretically perfect engine a reality. And eventually he managed to get financial backing for a project that, if presented by a less zealous prophet of industrial change, would have been laughed out of town for its impracticality and cost.

Rudolf Christian Karl Diesel was born in Paris in 1858. His parents were both Bavarians who had left an economically depressed southern Germany looking for work. His father, Theodor, was a craftsman in leather – the third in a line of Augsburg book-binders – and his mother, Elise, made a living teaching English and German. Diesel spent twenty-two of his fifty-five years in Paris, but his education was German; the company that gave him the chance to develop his engine was German; and as soon as he reached the minumum age of nineteen, he became a German citizen. He grew up in Paris, though, at a time when Paris was the most vibrant city in Europe. The Emperor Napoleon III was building the wide boulevards, the grand monuments and government buildings, the sewage system and infrastructure that gave Paris the expansive, confident character it still has today. Rudolf spent much of his spare time in the Conservatoire des Arts et Métiers, France's great museum of technology. He was preparing to attend the Ecole Primaire Supérieure when, in 1870, the Franco-Prussian war broke out. The war was accompanied by a growth in pan-German nationalism and consequently Bavaria came down on the side of the Prussians. As a result, all Germans became unwelcome in Paris.

The Diesel family left for London. In the long run this displacement was to Rudolf's advantage. If he had become a French citizen he would have been liable upon adulthood to nine years military service. (Avoiding

This pencil sketch was drawn by Diesel when he was 16 years old. A lifelong lover of art and music, he spoke three languages. And he got the highest marks in the history of his school, and did the same at his polytechnic. So I suppose you could call him a bit of an all-rounder.

Another example from Diesel's sketchbook. It is uncertain how well Diesel was acquainted with the subject of this sketch.

this personal catastrophe was one of his main motivations for taking German nationality when he did.) Rudolf didn't enjoy school in London, but was thrilled by the Science Museum. His parents couldn't afford to keep him in London, however, and decided instead to send him to stay with relatives in their native Augsburg. So at the age of twelve, Diesel was packed off across Europe on his own. The train journey took a gruelling eight days. On reaching Augsberg he was expecting to be taken in by his father's brother, but his uncle had no room, and Diesel was sent instead to a cousin, Betty Barnickel. She and her husband Christophe took in students to earn extra income, and Rudolf's uncle undertook to pay them the going rate of a thousand marks a year for Rudolf's board. He often couldn't scrape the money together, though, and so Rudolf ended up giving lessons in English and French in order to pay his rent. His own parents sent him very little. His stay with the Barnickels was not unpleasant, however. Betty Barnickel owed the Diesels a favour: she herself had been orphaned and brought up by Rudolf Diesel's grandfather, and so she treated Diesel well. And in one way the Barnickels were ideal foster parents for Diesel: Christophe Barnickel taught mathematics at the Augsburg Royal District Trade School, and so the interest in engineering that the young Diesel showed so clearly was well catered for.

Augsburg was a rather fortunate place for a budding young engineering genius to find himself. It had been a centre of trade and industry since the middle ages. When the first railways reached Bavaria in 1840, a heavy engineering industry developed that was to be essential to Diesel's career. By the age of fifteen, after two years in Augsburg, Diesel had decided to become an engineer. This meant one more year at the Royal Trades School, three years at the Augsburg Industrial School, followed by four years at Munich Polytechnic. His parents were very much against this lengthy course of education. Rudolf already owed money to the Barnickels for their hospitality, and the Diesel family themselves were living in poverty and needed all the help they could get from their children in order to get by.

Rudolf finished at the Trades School with very high marks. His family were now back in Paris and Rudolf went to visit them. Rudolf's father Theodor had always been a bit eccentric. His attitude to

By the end of the nineteenth century Diesel's home town of Augsburg was already a centre of heavy industry

education was influenced by his reading of the eighteenth-century philosopher, Jean Jacques Rousseau. It's sometimes said that a little learning can be a very dangerous thing, and this certainly seems to have been the case in Theodor's case. According to Theodor's interpretation of Rousseau's educational philosophy, in order to accustom and inure children to life's vicissitudes it is necessary to expose them to the experience of chance misfortune at an early age. And so, while out on country walks, Theodor would sneak up behind his children and trip them over, or else pick one of them up in the air quite suddenly and sling them into a ditch. And he was a severe disciplinarian: on one occasion, in an early expression of his fascination with all things mechanical, Rudolf dismantled a cuckoo clock, and left it in pieces. Theodor punished him for his curiosity by tying him to a chair while the family went for their regular country outing. It's difficult to know whether Rudolf would have been more upset at being treated in this brutal fashion, or pleased at missing the humiliating rituals of the walk. On another occasion Rudolf was punished for not telling the truth by being sent to school with a placard round his neck stating 'I am a liar'.

It is tempting to see in this strange early 'education' the source of Diesel's lifelong preoccupation with rationality. I'm not thinking so much of the effect of an early exposure to Enlightenment philosophy, as the effect of being exposed to the unpredictable and irrational behaviour of his father. Diesel originally called his engine the *rational heat engine*. He conceived it as part of a general project to reform industrial society according to rational principles. It was intended as a small, high-efficiency replacement for the steam engine. He wanted to bring mechanical power within the reach of small manufacturers, so enabling them to compete with the big steam-powered factories and helping to end the cruel social effects of urban industrialisation. Later in life he wrote a book called *Solidarism* in which he outlined the principles of a cooperative social system. Politically, Diesel seems to have believed that all it would take to reform society for the better was a good dose of rational behaviour, widely distributed. But 'rational heat engine' didn't stick. Diesel certainly wasn't too proud to cash in on the fruits of his philanthropic fervour, and realising that such a stodgy-sounding name wouldn't do him any favours when it came to shifting units, at the suggestion of his wife he re-christened his invention with his family name instead.

Despite his father's sometimes unpleasant behaviour, and despite their separation, Rudolf's relationship with his family was close. But when he visited them in Paris for the first time since leaving them in London, he must have been dismayed. His mother had grown excessively religious; his father was still unable to support his family and had become interested in spiritualism. While Rudolf was staying with them, his elder sister died. She was a talented musican and at the age of seventeen was already helping the family to pay its way. As well as increasing his determination to send Rudolf out to work, his daughter's death seems to have caused Theodor's interest in all things irrational to grow. He took up magnetic healing, but was unable to make any more money at that than at his leather work.

Rudolf wasn't swayed by his parents' entreaties. He decided he would be better off in Augsburg. Fortunately for him, the Barnickels took him back and the head of the Industrial School gave him money to continue his education. It was while he was studying there that he first saw the glass-sided Malayan firestick he was to use years later to demonstrate how his engine worked. He graduated in 1875 with the highest marks the school had ever awarded. A professor at Munich Polytechnic – who was also a commissioner for education in Bavaria – heard of Diesel's performance, and was sufficiently impressed to give him a scholarship to see him through the four-year course.

Diesel enjoyed studying at Munich. It was there, in the lectures of Dr Carl von Linde, that he was taught the theory of heat engines that gave him the inspiration for his future invention. Among other things, Von Linde was an expert in refrigeration, designing machines which were manufactured by the great Augsburg engineering company, Machinenfabrik Augsburg-Nürnburg, or MAN.

"Diesel had a vision of an engine that would sort out people's immediate social problems . . . All the engines we looked at ended up with a social agenda: the diesel is the only engine that started out with one."

THE MALAYAN FIRESTICK

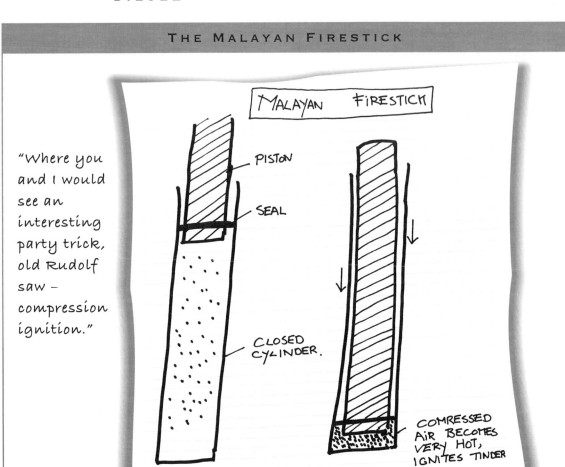

MALAYAN FIRESTICK

PISTON

SEAL

CLOSED CYLINDER.

COMRESSED AIR BECOMES VERY HOT, IGNITES TINDER

"Where you and I would see an interesting party trick, old Rudolf saw – compression ignition."

Force the plunger down quickly enough – before air has time to seep away around the plunger – and the increased pressure inside the cylinder produces an increase in temperature big enough to ignite the tinder at the bottom of the cylinder. Laurens Van der Post came upon one of these cylinder and piston contraptions while he was hacking his way through the Malaysian jungle. Malaysians used them for lighting fires with. Asking his sergeant what the devil the thing was, Van der Post was told, 'Basically, sir, it's a diesel engine.'

One of these devices was kept at the school Rudolf Diesel attended in Augsberg, and later in life Diesel borrowed it to use in a demonstration of why his engine didn't need spark plugs. But for Diesel compression ignition was only a useful spin-off. His interest in high compression was inspired by the theoretical increase in fuel efficiency that could be achieved.

This is where it gets really interesting – and really technical too. There's a good reason for sticking with it: by learning the theoretical background to the diesel engine, you're learning the principles that lie behind the success or failure of any heat engine whatsoever – even engines that haven't been invented yet. But if you're not in the mood for the heavy stuff, skip ahead to page 104.

Both Professor Linde's work on refrigeration and his lectures on heat engines were based on the work of the great engineering theorist, Nicholas Léonard Sadi Carnot. Carnot was born in 1796 to a brilliant family. His father, Lazare Carnot, was one of the five Directors of France – and a poet, a military engineer and a mathematician, who was also responsible for giving Napoleon Bonaparte his first military command. Sadi Carnot only lived to the age of thirty-six, and spent much of his short life as an artillery officer, but still found the time to study physics and economics. In 1824 he published his masterwork under the title *Reflections on the Motive Power of Fire and on Engines Fitted to Develop that Power.* Although the paper contains a good deal of mathematics, it is beautifully written in elegant, persuasive prose. Perhaps Carnot was aware of just how important his new insights were – or perhaps he just had a taste for a well-turned sentence. Either way, the paper had little success during his life, though its impact over the next fifty years, in engineering and physics, was immense.

Sadi Carnot

At the time Carnot was writing, nobody had given any convincing theoretical account of how engines go about converting heat energy into mechanical energy, and what the limits of this process are. Working on his own in his early twenties, Carnot managed to explain the whole business from first principles, and in doing so laid the foundations for the modern science of thermodynamics. It was because of Carnot's theoretical work that Rudolf Diesel could set out on the audacious project of replacing the steam engine – an engine of epoch-making efficiency and power that had taken over two hundred years to evolve – with something dreamed up on paper.

Carnot started with a completely general, abstract account of what a heat engine does. His account was valid for any heat engine whatsoever, no matter what fuel it uses to produce heat, and no matter what gas – or vaporized liquid – it uses as its 'working medium'. From

Refrigerators are heat engines run backwards – which is why Diesel could learn so much about engines from a man who specialised in fridges.

Refrigerators work a bit like sweating. We cool down when we sweat because the sweat absorbs latent heat of evaporation as it goes. Evaporation takes place in the cool part of the fridge (Latent heat of vaporization is absorbed); The resulting gas is compressed (work is being done on the gas, causing an adiabatic increase in temperature); The gas is allowed to cool to room temperature (heat escapes – including the latent heat of vaporization that was absorbed from the cold part of the fridge – and the gas liqifies); the liquid is pumped back into the cold part of the fridge, where it is allowed to evaporate once again. Refrigerators are really just mechanical sweating machines. Which makes you wonder about the wisdom of keeping food in them . . .

this starting point he could work up a theory of how an ideal heat engine would work.

Carnot argued that all heat engines are devices that exploit a flow of heat from hot to cold in order to convert some of that heat into useful work. The idea that engines are there to convert heat into work shouldn't raise any eyebrows. But the notion that it is a flow of heat from hot to cold that is essential, rather than heat itself, is less than obvious. Carnot supports his account by pointing out that even if the whole planet was as hot as the hottest boiler and steam could be made 'for free' and in abundance, it would not be possible to use the heat energy contained in that steam to drive an engine. There would be loads of heat energy around, but no heat flow. This matters because in a heat engine a gas does work by expanding. As the gas expands its pressure falls and, in accordance with Charles Law (the pressure of a gas is directly proportional to its temperature), its temperature falls too. But if the engine exhaust is no cooler than its intake, any fall in temperature in the gas will provoke a flow of heat to the gas from the atmosphere – and a consequent restoration of the original high pressure. If the gas cannot cool down, neither can it expand; if it doesn't expand, it doesn't do any work. In short, without a temperature gradient there can be no pressure gradient, and without a pressure gradient a gas cannot do any work.

Having abstracted the essence of all heat engines Carnot worked out what conditions an engine would have to fulfil if it were to be capable of transforming the greatest proportion of the heat energy contained in its fuel into useful work. That engine is the theoretically most thermally efficient engine there could be. It is called Carnot's ideal heat engine, and it operates on the Carnot cycle. Basically the ideal heat engine fulfills two conditions. The first is that it is fully reversible. The second is that it has the greatest possible difference between inlet temperature and exhaust temperature.

A reversible process is one which, if you run it backwards, brings you right back to where you started without having incurred any losses. Marriage, for instance, is seldom fully reversible, whereas rearranging the furniture is. A fully reversible engine is run backwards (as a heat-pump or refrigerator), pumping all the heat that has flowed from hot to cold back again from cold to hot will take no more work than was produced by the initial hot–cold flow. If any engine was more efficient than a fully reversible one, you could couple it up to a reversible engine and, running it as a heat pump, create energy out of thin air – which would be a contradiction of the laws of nature.

The reversible processes that lie at the heart of Carnot's ideal engine are adiabatic expansion and compression. A temperature change in the cylinder of an engine can be adiabatic in just the same way as the temperature changes that occur as air moves up and down in the earth's atmosphere. Starting with a volume of gas in a cylinder at a given temperature, if you do work on the gas by pushing the piston up the cylinder, and so reduce the volume of the gas and increase its pressure, the temperature of the gas will increase even though no heat is flowing into it. If the gas is then allowed to do work on the piston by pushing it back down the cylinder to its starting point, the temperature and pressure of the gas will return to their original levels. Ignoring friction and any heat that is absorbed by the cylinder itself, this adiabatic conversion of work into heat and heat back into work can be repeated over and over without loss of heat energy.

Adiabatic temperature changes are reversible because they convert heat energy into work energy and back again. So, since wherever a temperature difference exists, heat will flow from hot to cold (insulation never stops a heat flow, only slows it down), the only temperature differences that are allowed by Carnot in his ideal heat engine are adiabatic ones. This means that heat has to be taken in and out of the ideal heat engine isothermally. 'Isothermally' means taking place at constant temperature. To the ears of common-sense this sounds about as possible as demanding that someone should go from A to B without using any mode of transport. How on earth can you get heat to flow without a temperature difference to provoke that flow? Carnot's anwer is you do it by compensating for adiabatic temperature changes in the gas in such a way that heat flows without a net change

ADIABATIC LAPSE RATE

The fall in temperature as you go higher up in the atmosphere is adiabatic. That is to say no heat is transferred to or from the air; its heat energy is constant. The temperature difference between the top of a mountain and the valley below is due to the change in air pressure; gas cools down as its pressure falls. If you took the atmosphere away the temperature at sea level would be the same as the temperature on the top of Everest.

in temperature. Heat is introduced to the engine as the piston is moving down the cylinder. As the volume in the cylinder is increasing, there would be an adiabatic drop in temperature – except that heat is introduced at a rate which exactly compensates for that expected fall in temperature, so no net change in temperature occurs. Heat is removed from the engine in a similar way. As the piston is moving back up the cylinder, compressing the gas, the temperature of the gas would be increasing, but since heat is taken from the gas as it is compressed, there's no net increase in temperature. In both cases, adiabatic temperature changes are exactly compensated by heat flows into and out of the engine, so the actual temperature remains constant.

So Carnot's fully reversible engine remains at constant temperature, although heat is flowing through it and work is being taken out of it. Reversibility is an important idea. It was intuitively grasped by James Watt when he invented his separate condenser (see *Steam*). By spraying cold water into the working cylinder to condense the steam, Newcomen's engine created a big flow of heat before each and every power stroke, without increasing the amount of work done. By locating the hot and cold parts of the cycle in separate parts of the engine, Watt made the steam engine more thermally efficient – in Carnot's terms, closer to being fully reversible. Likewise, when Hornblower first experimented with the compound engine, and when Holt finally put the idea to good use almost a century later, Carnot's requirement of reversibility was being grasped in a practical way by men who could see that any waste of heat was a waste of fuel. But these reforms were truly practical applications of the ideal of reversibility. Carnot's criteria for full reversibility were utterly impractical. Which is no criticism of Carnot, who was engaged in the intellectual exercise of discovering the theoretical limits of heat engine efficiency. Unfortunately, Rudolf Diesel was so intoxicated with the perfect economy of Carnot's ideal that he tried to put the Carnot cycle into practice – lock, stock and barrel. He got into no end of trouble. More of that later.

The second criterion that Carnot's ideal engine had to fulfil was to have as great a difference in temperature between the heat *source* and the heat *sink* as possible. That is to say, the inlet temperature and the exhaust temperature should be as far apart as they can be. Now this is where the meat of Carnot's theory really lies – and where the seed of the success of Diesel's engine was sown. Carnot's argument was that thermal efficiency increases in direct proportion to the difference between hot and cold. A good way of understanding why this should be so is by going back to an old fashioned way of talking about the properties of gases. In 1662 Robert Boyle published his *New Experiments Physico-Mechanical touching the Spring of Air and its effects*. The 'spring' of air isn't a very familiar idea today, but it isn't hard to see that if you put air – or any other gas – in a cylinder, with a tight-fitting piston blocking off its exit, a spring is the perfect description for what you've constructed. Move the piston down the cylinder, causing the gas to expand below atmospheric pressure, and in effect you have stretched the spring; move the cylinder up the cylinder – squeezing the gas to greater than atmospheric pressure – and you have compressed it. And gas springs can be used instead of steel springs in some situations: the current Range Rover has a suspension system based upon gas springs.

Robert Boyle

But the opposite isn't true: you can't use steel springs to power an engine. This is because steel springs lack one startling property which only gas springs possess. Unlike a steel spring, the strength of a gas spring can be adjusted instantaneously, simply by adding or removing heat. Adding heat increases the gas pressure. And since pressure is force acting per unit area, this means that adding heat increases the force the gas is exerting against the piston. This will be true not just at the moment the heat is added, but all the time, until that extra heat is taken away again. Heat engines turn heat into work by exploiting this fact about gases: make a gas spring; compress it; add heat; let the spring expand, doing work as it does so. The spring will do more work in uncompressing itself than was done in compressing it because it is now stronger and exerts more force.

To get a bit mathematical about it, the work done by a spring is the product of the force it exerts multiplied by the distance through which it acts. So the greater the distance through which the spring is initially compressed, the greater the extra work yielded by allowing the spring to 'uncoil' after it has been strengthened by the addition of

THE MARCH OF HISTORY

It seems obvious in retrospect that Trevithick's high-pressure steam engine should be more efficient than Watt's low-pressure engine with separate condenser, and that Otto's compression stroke should be just what the internal combustion engine needed to turn it into a world-beater. But this is historical illusion. It is only in the light of Carnot's work that it is possible to understand why these two improvements were indeed improvements, and not sideways or even backward steps. Watt's steam engine heats steam to a moderate temperature then extracts every last drop of energy out of it by expanding it to a vacuum in a separate condenser. Compare that with Trevithick's high-pressure engine, which puts bags of energy into creating high-pressure steam, then simply exhausts it, still bursting with energy, out into the atmosphere. Surely you would have to say that, intuitively, Watt's engine seems likely to be the most efficient. The same could be said of Otto's four-stroke cycle. Everyone else was determined to achieve double-acting internal combustion engines. (Making every stroke a power stroke – surely you can't get any more efficient than that?) Otto didn't have a clue why, but when he pressurized the air/fuel mixture before igniting it the explosion became dramatically more energetic. The reason why no one else had thought of it before him was simple: there was no reason known to man to suppose it might happen.

Until Carnot, people knew that the high-pressure steam engine was a good idea only because, over time, they found it used less coal; and they knew that the Otto cycle worked because no internal combustion engine before Otto had been capable of producing more than three horsepower. Carnot's work explains both these improvements. The thermal efficiency of a heat engine and the amount of work it is capable of producing both increase in direct proportion to the difference between the temperature at the beginning of the power stroke and the temperature at the end. Applying this to the steam engine, it is clear that although Trevithick's engine exhausted steam at a far higher temperature and pressure than Watt's, it also took it in at a far higher temperature and pressure, and so the overall drop in temperature and pressure was greater in Trevithick's engine – and, consequently, so was the power output and the thermal efficiency.

As far as Otto's engine goes, while it is true that no more energy is put into the working gas by compressing it than could be taken out by just letting it expand again, once heat has been added the picture changes. The non-compression internal combustion engines that preceded Otto's four-stroke cycle could only extract the work done by the hot gas in expanding from just above atmospheric pressure back to atmospheric. But this is the bare minimum amount of work a gas that has had heat added to it can do in returning to atmospheric pressure. If you compress the gas first – before increasing its pressure by adding heat – you can turn the same amount of added pressure into extra work for the whole length of the expansion stroke.

heat. So if you want a bigger return on your investment of heat, compress the gas further before heating it up.

When Diesel was taught about Carnot's theories by Professor von Linde he became impassioned with the idea that he could make Carnot's ideal a reality. He made a note to himself on the margin of his lecture notes: *see whether it isn't possible to achieve the isotherm in practice.* As it happened, it wasn't. Carnot had never intended that his work should be used as a blueprint for an actual engine. In his treatise he stressed that it was intended only as method of analysis – a benchmark against which the performance of actual engines could be compared, understood and criticised. And he took care to stress that other aspects of an engine's performance – such as reliability, economy of manufacture, complexity, safety, size and weight – may well be more important in practice than thermal efficiency. The job of the engineer is to balance competing factors and come up with the best practical solution in the circumstances. But Diesel doesn't seem to have come across this part of Carnot's work. Or else, if he did, he took no notice, because his first patent was nothing more or less than an attempt to realise Carnot's ideal cycle in an internal combustion engine. Even so, his effort to achieve this impossible dream did result in the development of a completely new engine – one of the most influential there has ever been.

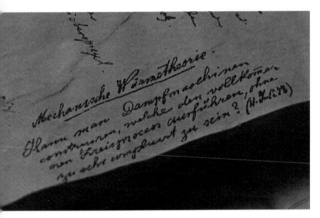

Where it all began: Diesel's famous margin note.

That was the difficult bit. If you've skipped the last seven pages, rejoin the text here.

In his lectures Professor Linde explained that the steam engine had a theoretical thermal efficiency of only six to ten per cent. This means the steam engine converts at most one tenth of the energy produced by the burning of coal into useful work. Diesel seems to have taken moral offence at this statistic. Years later, when he was married and had children, he would still point out steam engines to his family and go on about how wasteful they were of the world's resources. The children grew up with the impression that the steam engine had done them some unspecified but serious harm. Oddly enough though, Diesel was a great admirer of James Watt, and always spoke of him with awe and reverence. Perhaps this is because, of all the great steam engineers, Watt was closest to the academic establishment. Legend had it that in 1760 when Watt heard about Joseph Black's discovery of the latent heat of vaporization, he suddenly understood what was wrong with Newcomen's engine and rushed off to invent the separate condenser. This almost certainly isn't true, but it could explain why Diesel identified with Watt so closely.

It was a little while before Diesel actually started working on building his ideal heat engine. In the summer of 1879, when he was about to take his finals, he fell seriously ill with typhoid fever. He made a good recovery, but missed his examinations. Professor Linde got him a spell of work experience with the Swiss engineering firm, Sulzers. At Sulzers Diesel got his only shop-floor work experience – spending four weeks as an apprentice on a filing bench. He returned to Munich early in 1880 and was given an oral examination, which he passed with the highest grades the polytechnic had awarded in its twelve-year history. Needing a French speaker to look after his business in Paris, Professor Linde gave the job to Diesel, and after only one year working in Paris, Diesel became Linde's manager there. So, by the age of twenty-three, he was making very good money. Like his father before him, he met a German girl in Paris and married her. Then, having settled his home affairs to his satisfaction, Diesel turned his attention to developing a revolutionary new heat engine – like you do.

Before Diesel got around to working on his internal combustion engine he spent several years attempting to reform the steam engine. Steam engines are external combustion engines. External combustion just means that combustion takes place outside the engine's working medium. Though steam has powered the only really successful external combustion engines, it isn't the only possible working medium. Diesel reasoned that one of the chief inefficiencies of the steam engine was the necessity of heating water up to 100 C to vaporize it before it can be used to power the engine. On expansion, as soon as the steam reaches 100 C again it condenses. So, by using water as the working medium in an external combustion engine you're using up a whole lot of heat just in turning your working medium from a liquid into a gas. None of that heat gets turned into work, because as soon as the steam temperature falls back to 100 C it becomes useless.

Diesel was used to working with ammonia in the course of his employment with the refrigerator company. He knew that ammonia vaporizes at a far lower temperature than water, and so he reasoned that if he could design an external combustion engine that ran on ammonia instead of water, the working medium would be raised to a higher temperature and pressure with exactly the same input of heat. The idea was sound in principle, but ammonia is a difficult substance to handle and, despite several years' work, nothing came of his efforts. But according to Diesel himself, the work he did on his ammonia engine laid the foundations for his subsequent invention of a high-pressure compression ignition engine. While working on the ammonia engine he made a systematic study of the behaviour of vaporized

THE DIESEL CYCLE

The first stroke draws fresh air into the cylinder. During the second stroke the air is compressed to extremely high pressure. The increase in pressure causes a dramatic increase in temperature. At the top of the compression stroke liquid fuel is sprayed into the cylinder. The high temperature of the air causes the droplets to vaporize and then ignite. So the air in the cylinder is heated even further, expanding as it does so and forcing the piston down the cylinder (the third stroke). In the fourth stroke the hot gases are expelled from the cylinder.

The direct injection of fuel into the combustion chamber is a way of enabling much higher rates of compression to be achieved without having to worry about pre-ignition.

An incidental pay-off is that the diesel engine doesn't need spark plugs or any of their electrical paraphernalia, and is therefore that much more reliable.

liquids, and learned that they behave in exactly the same way as 'normal' gases, like air. This gave him the idea of treating air not only as a source of oxygen for an internal combustion engine, but as the working medium itself. Combustion would just be the means of heating the air. It also gave him the idea that the best way to exploit the extra efficiency, which Carnot had shown was the result of a high-temperature differential between induction and exhaust, was not to add more heat to the working medium, but rather to compress it further before adding the heat.

This last idea was, indeed, a momentous realisation. Diesel's critics were to argue that it constituted no discovery at all: Otto's four-stroke engine utilised exactly the same principle in practice. But in fact, most engineers at that time – Sir Dugald Clerk, designer of the first successful two-stroke cycle engine, among them – were arguing that the losses due to friction would far outweigh the benefits in efficiency produced by increasing compression ratios. So, far from ripping off Otto's insights, Diesel was flying in the face of accepted wisdom.

So Diesel turned his attention to internal combustion as a means of realising Carnot's ideal engine in practice. He applied for a patent on a high-compression internal combustion engine in 1892. The engine he described was nothing more or less than Carnot's ideal heat engine incarnated in the form of an internal combustion engine. A

ENGINES THAT DIDN'T TURN THE WORLD
THE AIR ENGINE

Carnot recommended that air was the most suitable working medium for heat engines. In a way, all internal combustion engines are partial air engines, as air is a major component in their working medium, but the composition of the air is transformed when combustion takes place and the oxygen combines with the fuel. Rudolf Diesel's initial design for a high-compression engine, on the other hand, used so much air in relation to the amount of fuel it burned that it was almost a pure air engine. Combustion was just a means of imparting heat to the air in the cylinder. But this wasn't the first stab at using air as the working medium for a heat engine. The air engine has a long history. And it is a curious fact about this history that the two great proponents of the air engine were both vicars.

The idea of an air engine was patented back in 1759 by the Reverend Henry Wood. He envisaged a Newcomen-type engine that used air in place of steam. During the nineteenth century several air engines were built, though not in consequence of Henry Wood's patent, which was rather vague and outlined nothing more than the idea and its advantages. The best of the nineteenth-century products was patented by the Reverend Robert Stirling in 1816. The Stirling engine has received renewed interest in recent years because of its possibilities as a low-emission engine.

cylinder full of pure air would be compressed to 250 atmospheres (an astonishing 3675 psi – totally impractical at a time when very high pressure meant 250 psi). The first stage of the compression stroke would be isothermal. Increase in temperature would be avoided by spraying cold water inside the cylinder. The water would absorb just enough heat to compensate for the adiabatic temperature rise due to compression. The second part of the compression stroke would be adiabatic, causing the temperature inside the cylinder to rise – to around 900 centigrade. As the piston started back down the cylinder, fuel would be gradually introduced into the cylinder, igniting as it made contact with the hot air. By introducing the fuel to the cylinder at such a rate that the addition of heat would be offset by the reduction in temperature due to adiabatic expansion of the gas, isothermal combustion would also be achieved. In the second part of the power stroke the air would be expanded adiabatically all the way to atmospheric pressure. Because combustion would be isothermal, Diesel reasoned, cooling of the cylinder would not be necessary.

In retrospect it is surprising Diesel wasn't laughed out of town for his patent. On paper his engine was capable of thermal efficiencies in the region of seventy to eighty per cent – a seven-or eightfold increase on the ten per cent figure for the steam engine that had so outraged him as a student. Sounds pretty good. But Diesel's expectations of what could be achieved in practice are less suggestive of a healthy optimism than an overactive fantasy life. Diesel was so hypnotised by thermal efficiencies that he failed to take any other kind of efficiency into account. It's worth remembering here that

Diesel's original patent

thermal efficiency isn't the same as fuel efficiency. Fuel efficiency is a common-sense notion, a simple measure of how much fuel an engine uses to perform a given amount of work – the sort of a practical measurement a businessman would be interested in when calculating whether a piece of machinery is a good investment or not. Whereas thermal efficiency is a much more rarified notion, a measure of the proportion of the thermal energy released by the fuel during combustion that is converted into useful work. It's a theoretical measure that describes only one aspect of an engine's performance. Once mechanical efficiency is taken into account, the more thermally efficient engine might well turn out to be less fuel-efficient than a less thermally efficient one. So heading all out for thermal efficiency isn't necessarily as good an idea as it sounds at first. Mechanical efficiencies had been well understood since Isaac Newton's day. It is amazing that such a brilliant man could leave such an important aspect of any engine's performance out of his calculations.

In fact, allowing for mechanical efficiency, or lack of it, it might even turn out that the most thermally efficient engine of all would not even generate enough power to overcome its own friction. And that's exactly how it did turn out. Diesel discovered this only after he exposed his plans to the criticism of his peers. Isothermal combustion and exhaust are simply a guarantee that no heat whatever is going to waste. The downside of 'achieving the isotherm in practice' is that while it may well be the case that no heat is being wasted, it might equally be the case that precious little work is being produced. And while Carnot was absolutely correct to argue that the most thermally efficient engine possible would have the greatest temperature difference between the heat intake and exhaust, at a certain point any further increase in an engine's compression ratio produces so much extra friction that more work is absorbed in the compression stroke than is liberated in the power stroke. Similar disadvantages arise when you try to extract every last drop of work from the gas, expanding it all the way to atmospheric pressure. When the pressure differential between the gas in the cylinder and the atmosphere drops below a certain level, although the piston can still do work, so little power is produced by the expansion that more work is lost to friction than is done by expanding the gas. In short, the theoretical thermal efficiency of Diesel's engine was terrific – and so it should be, considering it was designed as a practical realization of the Carnot cycle – but even if an engine could have been built to withstand such enormous pressure, it would not have produced enough power to overcome its own friction. Diesel was like a man who claims to have perfect teeth but, when asked if he would mind showing them, proudly gets them out of a box and reveals how little decay they have.

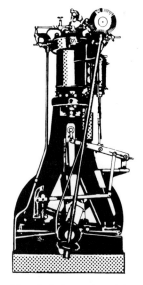

Diesel's first experimental engine (above) achieved an efficiency of 16.6 per cent.

"Rudolf Diesel wanted to produce an engine that was so efficient that it didn't even need cooling. Well, many diesel engines can run at half revs all day without producing a lot of heat. You see them at fairgrounds running with their radiator caps off. A little boy goes round with a jug of water topping them up. It's a sure sign that they're running cool."

Despite their impracticality, Diesel's ideas received much public interest. Diesel took the criticisms on board and over the next year gradually adapted his rational heat engine to the requirements of mechanical efficiency. He was forced to abandon much of the content of his original patent, increasing the engine's power output at the cost of reduced thermal efficiency. Isothermal compression was the first thing to go. The injection and exhaust of water from the cylinder on every compression stroke was clearly going to absorb more power than its efficiency saving was worth. The ultra-high compression pressures were reduced to 40 atmospheres – relatively modest but still way in excess of anything that had been achieved in any previous engine. Constant temperature combustion was replaced by the more powerful but less thermally efficient constant pressure combustion, in which the fuel is introduced into the cylinder slightly more quickly and the cylinder temperature is allowed to rise a little. And the idea of expanding the working gas all the way down to atmospheric pressure went by the wayside as well, in favour of stopping expansion early and exhausting the gas when it was still above atmospheric pressure.

So by November 1893 Diesel had given his plans a thorough overhaul – and for the first time his projected engine looked like a practical proposal. But Diesel could not admit in public that his rethink amounted to more than tinkering with the initial theory. This wasn't pride – at least, it wasn't *only* pride; his original patent specified an engine with isothermal combustion, high compression and compression ignition, all of a bundle. If these characteristics had been listed separately Diesel could have ditched the bad eggs and kept the remainder, but because they were linked, German patent law meant that owning up to having abandoned some of those characteristics would have left Diesel with no legal rights over the remainder. So Diesel was committed to distorting the nature of his work and his invention, and spent much of the rest of his life in needless arguments over its originality. The need to disguise his tracks was so great that it isn't quite clear whether Diesel actually believed his own falsified account or not.

After approaching several firms – gradually reducing his proposed compression ratio until it fell within the realms of the

The single cylinder of Diesel's first working engine was built like a cannon. It had to be to withstand the massively high pressures Diesel demanded. It was shaped a bit like a cannon too, with a bore of 150mm and a lengthy 400mm stroke. The crude oil Diesel had intended to use as fuel wouldn't flow, so instead Diesel employed lamp oil. After some work, a compression of almost 500 pounds per square inch was achieved. If Diesel had ever read Carnot's warning about the dangers of trying to realise all the motive force of combustibles, it probably flashed through his mind when he first turned over this engine: the violence of the combustion was so great that the pressure gauge exploded and bits of glass shot in every direction. Amazingly, nobody was hurt.

possible – in 1893 Diesel finally persuaded Maschinenfabrik Augsburg Nürnburg to take on his project. They were attracted by his suggestion that the engine might be developed in a form which would allow the use of coal dust as fuel. An almost useless by-product of the mining industry, coal dust was the cheapest of all fuels. An engine that could utilise such an inexpensive fuel would have been a phenomenal success. In fact the suggestion was impractical, and it seems likely Diesel already knew as much. Coal dust doesn't flow, so feeding it into even a slow-speed engine in accurate quantities is extremely difficult. But MAN took the bait, and the first engine was built that year.

"This is the key to Diesel's design: efficiency – ruthless efficiency . . ."

FUEL FOR THE FIRE

Rudolf Diesel never took to petrol. Perhaps he didn't like the smell. More likely it was because refining petroleum from the heavier components of oil costs good money. For his rational heat engine Diesel wanted to utilise the cheapest fuel he could. Because he used fuel injection rather than a carburettor system, his fuel didn't need to vaporize at atmospheric temperatures. Injected as a liquid, petroleum oil would vaporize almost instantly in the great heat of the highly compressed air. He wasn't the first person to exploit the low cost and relative safety of such fuel. Petrol was considered to be very dangerous in the nineteenth century and its use was often strictly controlled. So several types of oil engines were designed and built. Some of them were commercially successful. One such was the Ackroyd Stuart engine.

Herbert Ackroyd Stuart was born in 1864. He was the son of a Scotsman born in Yorkshire, so he will have known a thing or two about economy. His engine was sufficiently like Diesel's engine for him to kick up a fairly convincing fuss over Diesel's patent rights, but in reality the similarities between the two engines were only skin deep. True, Stuart's engine didn't use electric ignition, but it didn't use compression ignition either. Vaporization and ignition of the oil was achieved by means of a hot bulb leading off the combustion chamber. To start the engine the bulb was heated by means of a torch. This took between ten and twenty minutes. Once achieved, operating temperature was maintained by the heat produced by the engine itself. The principle was the same as in an ageing petrol engine which gets so hot inside that it continues to run on even after the ignition is switched off. The Stuart Ackroyd engine was very simple to run and reliable, so it became popular in situations where an engine was needed that would run without anyone being in attendance – it sold particularly well to farmers.

Diesel's original experimental engine

After many modifications the engine ran on its own for the first time early in 1894. When Diesel came home that day he embraced his wife and burst into tears. He took her to the workshop and had her set the engine running herself. But it wasn't until 1895, when a third version was built, that the engine became a reliable runner, producing 23 brake horsepower and using only half the fuel of the most efficient engines of the day. At 16 per cent, the new engine's thermal efficiency was double that of most contemporary steam engines and half as good again as the best petrol engines. Soon figures of over 20 per cent were achieved. In 1896 a production engine was designed and on 17 February 1897 a professor from Munich University performed 'acceptance' tests on the engine. His tests showed that it had a thermal efficiency of 26 per cent – twice as efficient as the best petrol engines of the day, and almost four times as efficient as ordinary steam engines. He declared Diesel's engine to be the most economical in the world, and definitely the engine of the future. Diesel instantly became famous, and it wasn't long afterwards that he became rich. A good job too, because his father had been borrowing money on the expectation of his son's imminent victory over the dark forces of steam.

But after Diesel had back-tracked so thoroughly on his initial patent, what was left of his original idea? Critics argued that there was nothing left – that Diesel was just snatching a variety of existing ideas and treating them as his own. Despite achieving fame and fortune, Diesel felt hounded for the rest of his life by accusations that he was not really responsible for the great advance that the diesel engine represented. It is true that the form the engine eventually took shadowed the Carnot cycle far less closely than Diesel had intended, but it did embody certain aspects of the cycle, and that was why it made such drastic improvements in thermal efficiency over the steam engine and the petrol engine. For one thing, Diesel's engine had a more gradual fuel combustion than Otto's petrol engine – and consequently a more even cylinder temperature, a smaller heat loss to the atmosphere and a greater degree of reversibility. But more significantly, the all important temperature difference between intake and exhaust was greatly increased over the Otto cycle.

Critics could argue that this wasn't a difference in kind to the Otto cycle, but only a difference in degree; the compression ratio of petrol engines may well have been gradually increased over time. But it was only by compressing pure air and injecting fuel afterwards that such compression ratios could be achieved. In the Otto cycle the compression ratio is always going to be limited by the combustion temperature of the fuel, because the Otto cycle draws fuel and air ready-mixed into the cylinder. During the compression stroke, as soon as the mixture reaches the combustion temperature of the fuel it will start to burn. If the compression ratio is very high this happens before the piston has reached the top of its stroke, so the engine is in effect working against itself and suffers a dramatic loss in power. That is why petrol engines are limited to compression ratios in single figures. As the result of chemical additives to petrol, Otto cycle compression ratios have increased over the years – but they're still not in the diesel league. Compressing pure air means that pre-ignition is never a problem and the only effective limit to the compression ratio is friction.

Whether or not someone else would ever have eventually spotted the practical possibilities offered by Carnot's theory is not a question that needs to be answered, since Diesel spotted them first. What is important is that nobody could ever have come up with the high-pressure compression-ignition engine as a result of trial and error experimentation in the workshop. The air in the cylinder of a diesel engine is under such tremendous pressures that the difficulties of developing a fuel injection system capable of forcing a measured amount of fuel into the cylinder were immense. In fact, no really satisfactory system was developed until the 1920s. So it is hardly possible that someone could have stumbled upon the benefits of high-pressure fuel injection in the way Otto happened upon his compression stroke while tinkering in his workshop. Indeed, it was years until Diesel himself saw any proof that what he had calculated would work on paper was going to work in practice. Who would have seen such a development process through without a firm belief in the theoretical justifications for their project? That's where Diesel's brilliance and originality lay.

Diesel was also the first person to utilise compression ignition in a heat engine. Compression ignition gives the diesel engine a whole

"There's no carburettor on a diesel engine. The fuel is pushed straight into the cylinder. Diesel injectors are horrendously complicated. They have to be to push fuel out at a couple of thousand pounds per square inch. If they didn't it wouldn't come out at all, the pressure in the cylinder is so high. Even guys who work on diesels will tell you that if anything goes wrong with the injector you send it back to the manufacturers. You do not unscrew anything."

THE GARDNER DIESEL

Gardner took the intrinsic reliability of the diesel engine and turned it into a virtue in itself. Gardner diesels are manufactured to extraordinarily high tolerances. For anyone who

The Lizard Lighthouse in Cornwall has been powered by diesel engines built by the Manchester firm of Gardner's for fifty years

has a feeling for machinery, their quality is inspirational. That's why for more than half a century they were the first choice where reliability was the main criterion. The lightships that used to dot the British coastline were powered by them, and many lighthouses used them too, although the Lizard Lighthouse is the last to do so today.

Gardner diesels were also favoured for small fishing vessels: when you're bobbing about in the North Sea in heavy weather you don't want to have to worry about whether your engine is going to start or not. As the size of fishing vessels has increased, the importance of the Gardner in the industry has lessened, though they are still used as auxilliary engines in case the larger, more advanced but less reliable main engines cause trouble.

Stories abound about Gardners being picked up from scrap heaps, cleaned up, refuelled and started right up again. There's nothing much that can go wrong with a diesel engine, and if they're manufactured with longevity in mind they really do just run and run.

"They made half a dozen special engines in the thirties and put them in Lagondas – to prove that they weren't just big coarse industrial motors."

THE GARDNER DIESEL

"The Gardner engine is a thing of rare beauty. . ."

". . . an engine designed with meticulous attention to detail and built by men whose workmanship stands the test of time."

Most of the generators that produce the electricity to run the rides at funfairs are built around old Gardner diesels that have already had one or even several former lives. Although the Gardner isn't the most thermally efficient of diesel engines, building an engine so well that it hasn't been scrapped after twenty years – or even after fifty – is a rational approach to engine manufacture that Rudolf Diesel would surely have approved of.

heap of advantages apart from fuel efficiency. Making spark plugs spark at the right time requires a battery, an alternator to keep the battery charged, a coil to raise a high enough voltage to give a good spark, and a distributor to send current to the right spark plug at the right time. The timing of the sparks has to be kept well adjusted if the engine is to run efficiently, and if the engine is to reach its maximum potential throughout its range of speeds, some way of advancing the timing as engine speed increases also has to be devised. Together these components form by far the most vulnerable part of a petrol engine. They're sensitive to dirt, damp, accidental damage and extremes of temperature. Doing away with the need for them abolishes many of the most common reasons for engine breakdown, making the diesel engine significantly more reliable than the petrol engine.

"Underneath his house, Diesel built a whole warren of connected corridors, simply so that his children could bicycle around if the weather was bad . . ."

Representatives of engineering companies from all around the industrialised world flocked to Augsburg to secure licensing agreements with Diesel. In return for their generous and patient support of his development work, Diesel had already given German rights to MAN, but rights for the rest of the world remained his to sell. By the turn of the century over fifty licences and sub-licences had been granted – even though the engine was still in a fairly crude state of development. But the success of the project that had occupied him for much of his adult life seemed to bring Diesel little happiness. His severe headaches continued. The only solution appeared to be complete rest, and Diesel was obliged to spend time in a sanatorium. It's not clear what the modern diagnosis of his problem would be. He had fantasies of vying with the likes of Rockerfeller and Nobel for power and influence, but, lacking in business acumen, only succeeded in wasting a lot of money. He squandered a good deal on fancy living as well, spending a million marks on building a grand house in Munich for himself and his family.

As his fame increased he went on lecture tours, tirelessly promoting his engine, the development and performance of which continued to improve, and Diesel became friends with some of the foremost technologists of the day. But even after his patents expired in 1907, engineers all around the world still attacked him on the grounds that he had invented nothing new and had no right to put his name to his engine. This bothered him a lot more than it should have done – perhaps because, as a result of his delicate patent situation, he had not been completely honest about the genesis of the engine himself.

By the beginning of 1913, Diesel had lost more than ten million marks through ill-advised speculation and he now had gout, insomnia and a heart complaint to add to his troubles. He started to read the philosophical works of the great pessimist Arthur Schopenhauer. On 29 September he boarded the steamship *Dresden* in Antwerp, bound for Harwich. He was to lay the foundation stone of a new diesel engine factory in Ipswich, and pay a visit to his friend Sir Charles Parsons, inventor of the steam turbine. But he never arrived. His bed was not slept in and his hat and coat were found on deck, neatly folded. Diesel's body was dragged out of the sea by fishermen a few weeks later. In accordance with marine tradition, they stripped it of possessions and returned the corpse to the sea. The 55-year-old Diesel had confided his financial problems in nobody. The nearest he had come to discussing his despair was in a letter to his wife. But he posted the letter to the wrong address and was dead by the time she received it.

Despite his success, Diesel became more and more subject to depression as he grew older. Reading Schopenhauer almost certainly didn't help.

At the time of Diesel's death diesel engines cost about three times as much as petrol engines of equivalent power. But their superior fuel efficiency guaranteed continuing interest. The goal was to improve the power-to-weight ratio of what was still at this time an extremely bulky and heavy power plant. By 1914 Sulzer had developed an effective and powerful two-stroke diesel for marine use. Diesel engines doubled the effective range of submarines, as well as avoiding the problems of petrol fumes in an air-tight environment. At the beginning of the First World War, the success of German diesel-powered U-boats prompted an international scramble to develop diesels with better and better power-to-weight ratios for military applications. The German engineer Hugo Junkers even succeeded in developing a diesel engine to power aircraft, and in 1938 a plane powered by this engine took the world long distance flying record.

One of the main limitations of the early diesel engine was the fuel injection system. Because at the moment of combustion the air in the

One of the peculiarities of the diesel engine is that, compared to the petrol engine, it has quite a narrow power band. This can only be compensated for by having plenty of gears. This Mack truck has sixteen – eight separate gears and a two-speed back axle.

cylinder of a diesel engine is under extreme pressure, getting a measured amount of fuel into the cylinder is no easy business. Although he was unhappy with the system, Diesel had had to make do with a pressurized air injection system. This required a compressor to be run off the engine, increasing the engine's size and weight considerably. The first person to come up with an acceptable lightweight replacement for the compressor system was an engineer by the name of Prosper L'Orange. L'Orange worked for the motorcar manufacturer Karl Benz. Between 1909 and 1919 he developed the first successful system of 'solid' fuel injection. The first production diesel engine without a compressor was offered for sale in 1920, and a diesel-powered truck followed in 1924. It wasn't fast but, load for load it would travel the same distance at the same speed as a petrol-powered truck for less than one fifth of the cost in fuel.

Diesel power took off rapidly on this side of the Atlantic. When London Transport tested diesel-powered buses they found they only consumed half the fuel of the petrol-driven models, and immediately set about converting their fleet to diesel. Fuel was considerably cheaper too – even after the government upped fuel tax on diesel by a factor of seven. The story was the same all over Europe. But when nine-tenths of all light trucks in Germany were already powered by diesels, there were fewer than 1500 diesel-powered trucks in the whole of the United States.

There are several possible reasons for the slow acceptance of the new engine in the USA. The United States is and was a country awash with cheap fuel, and blessed with the God-given right to burn it. Extracting every last bit of work from the thermal energy contained in your fuel is hardly the American way, after all. Some people have also suggested that anti-German prejudice fouled the pitch for diesel proponents for many years, and it is true that when the engine finally took off it was in the mid west of the country where the German influence was greatest. Another possible reason for the limited enthusiasm shown for Diesel's invention was the American style of engineering. Compared to the European tradition, the tradition of American engineering is very much more low-tech. Highly skilled engineers were not easy to come by in the States in the early part of the century, and the standard of machining and casting demanded by the

"The automotive diesel was developed in 1923 . . . but it was a long time before it broke the monopoly of Ford's cheap and cheerful petrol-driven car."

The Caterpillar no. 2 tractor (left and below) – the second diesel tractor to be manufactured and the first to be sold in the United States. Caterpillar no.1 was kept back as a demonstration model.

"It was a difficult market for diesel to get in to. Why would anyone bother with this newfangled diesel thingy when you could run a big petrol engine like this one for sweeties? But the Depression was to change all that . . .

. . . and by the end of the 1930s half the farming in the United States was done by diesel."

diesel engine was much greater than that required by the petrol engine.

But that all changed in the 1930s – probably as a result of the Great Depression. Food prices dropped so low that farmers could hardly make a profit on their crops. In 1931 the Caterpillar tractor company offered their first diesel-powered tractor for sale. It was thirty per cent more expensive than their petrol-powered model, but it sold ten times as well. All of a sudden an improvement in the economy that for years had been too marginal to matter became the difference between turning a profit and going to the wall. Caterpillar sold ten thousand tractors in the next few years and other manufacturers of farm equipment soon introduced diesel models of their own.

So Americans came to appreciate the advantages of a very European invention. And eventually they took it up with enthusiasm: it was in the United States of America that steam locos were first

CATHEDRAL ENGINES

"This is about as big as diesel engines get. These 3-storey monsters are called cathedrals – and it's not hard to see why."

"It's an astonishing thing to climb up an engine and be out of breath between the crankshaft and the cylinder head."

"It's a straight 8 two-stroke supercharged diesel and it produces about 24,000 horsepower . . ."

Almost all commercial vessels are powered by diesels now. Most marine 'cathedral' diesels run on a two-stroke cycle. When used in a diesel, the two-stroke has all its usual power-to-weight advantages, plus the added bonus of far greater cleanliness. That's because all diesels have direct fuel injection, so fuel doesn't have to be introduced into the cylinder while the exhaust gases are still being cleared. This ship's engine runs at around the same ponderous 80 r.p.m. as Diesel's very first functioning engine. It is so powerful it has no need of gears.

". . . That's absolutely state-of-the-art diesel technology. This engine is more than 50 per cent efficient – that's three and a half times more efficient than any steam engine ever was."

CATHEDRAL ENGINES

"They carry a spare cylinder and a spare piston that are so big and heavy that they have a special overhead crane to move them around."

"You always expect when things get enormous that they'll do all sorts of clever things, but no: the bolts holding the crankcase to the soleplate – the big mounting plate – were eight inches across with thread on them you could put your hands into. I asked one of the engineers how did they do them up and he said, 'Many men. Many men on long bar.' And how tight do they go? 'Tight as you can get them.' In other words, it's one of the most efficient internal combustion engines in the world, and they've got a whole control panel covered in electronic monitoring devices, but working on the actual engine is not a million miles away from tinkering with a truck in your garage."

"There's more turning power in this shaft than there is in almost any other man-made object in the world. The noise that all this machinery makes is quite unlike anything else . . . It's more a feeling than a noise. It comes right up through your feet. And it's kind of scary and terribly exciting at the same time."

driven out by the now ubiquitous diesel. Seeing those inefficient steam trains taken out of commission and replaced by far more thermally efficient diesels would probably have had Diesel dancing in his grave. But the underlying logic of the decision wasn't entirely to do with thermal efficiency. There are many other arguments for diesel locomotives in preference to steam. Diesel locomotives can be started up in an instant, whereas steam locos have to be woken early in the morning and kept constantly tended throughout the day. On top of that, to run a railway service it is necessary to have locomotives on standby to replace breakdowns, and the working locomotives inevitably end up standing idle for at least a part of the day, simply because it isn't possible to keep all the locomotives active all the time. So even if steam engines were more fuel-efficient than diesels, they would probably end up wasting fuel because of the practical requirements of running a railway. Steam engines are also very labour-intensive to operate. The fire in the firebox has to be started afresh each day, constantly stoked throughout the day – usually by hand – and the firebox has to be emptied and cleaned at the end of the day. A steam train cannot be operated with fewer than two people, and regular stops to take on water and coal mean a slower service for the customers. And steam engines are dirty, producing an unacceptably large amount of smoke, often falling foul of clean air legislation wherever this was implemented.

"When people started to get proper wages steam became problematic. There are a lot of dirty, poorly paid jobs associated with steam. On the big liners the stokers worked twelve hours on and twelve hours off. A terrible job. No one would do it if they had a better offer."

Diesel engines, on the other hand, can be started in minutes and turned off again as soon as they are no longer needed. They can be operated by a driver working alone, they do not need to make stops other than to wait for signals and to pick up passengers, and they are relatively clean. They have the disadvantage over steam of increased mechanical complexity. Apart from the fuel injection, the main reason for this added complexity is that like all internal combustion engines, diesel engines only have a good power output over a certain range of engine speeds. This means the engine has to be kept turning within that range of speeds. The gearbox is there to gear the engine speed up or down, altering the number of revolutions of the driving wheels that are produced by a single revolution of the crankshaft in order to make a wide range of speed over the ground possible without loss of engine power. The power produced by a steam engine, on the other hand, is only determined by the pressure of the steam in the boiler. So steam locomotives have direct drive between pistons and wheels and no need of gearing.

This added complexity means that diesel engines are more expensive to manufacture and require a more sophisticated standard of maintainance. Even so, when wage bills are taken into account diesels still come out cheaper. But what happened in the USA in the fifties when steam locos were replaced by diesel wasn't so much a reflection of these arguments, as a response to the political situation of the day.

Transportation is always of central economic importance. Without transport, trade and industry will grind to a halt. Look at any map and you'll see that all over the world, the pattern of settlement and economic development has been influenced by transportation. Before the railways, water carriage was the most economically significant method. The major cities grew up on navigable rivers and by sea ports. The early flourishing of civilisation in the eastern Mediterranean can be put down to the possibility of trade and communications between the many ports which nestled around the same, relatively small body of water. Likewise, Britain's success as a trading nation and its accelerated economic development was due to its access to the sea (nowhere in the UK is more than seventy miles inland). So in such a vast country as the United States, internal transportation is always going to be crucial. By the middle of the twentieth century, America was home to a massive one-third of the world's total mileage of railway track. Clearly, railways had become central to the effective functioning of the US economy.

In the fifties that economy was flourishing. Work was easy to come by and the unions had grown strong. The mining unions were particularly active, and because the railways were dependent on steam engines for motive power, and steam engines were dependent on coal, by cutting off coal supplies to the railways, the mining unions were able to hold the entire US economy to ransom, and they used

"The great attraction about the diesel is that it has the most fantastic tractive power at low revs, which makes it absolutely ideal for pulling huge loads long distances."

their power effectively. The rapid changeover from steam to diesel was very much a response to that power. It said more about the economic advantages of the relatively low levels of union disruption in the petroleum industry than about the relative efficiency of diesel and steam.

Britain followed suit in the 1960s. And you might have thought that by then the debate between steam and diesel would have been boiled down to its bare economic bones, but policy had a large part to play in that decision too. Although many people will tell you that working with steam locomotives gave them a wonderfully satisfying career – and many camera-clicking steam enthusiasts envy them for it – when it comes to rolling your sleeves up and getting your hands dirty, most normal people experience a degree of reluctance. Working with steam is hard graft, and it's dirty. The temperature inside the cab of a steam loco is often uncomfortably high. So in the sixties when unemployment levels fell and it was easy enough for young people to find clean, physically undemanding work, the railways found it difficult to recruit enough school leavers to keep the service running.

This was a serious problem, added to which was the problem of a great shortage of suitable coal. Steam locos require fist-sized lumps if they are to function effectively. This is because the effect of the forced draught produced by directing exhaust steam up the chimney is so powerful that smaller pieces tend to get sucked through the boiler and spat straight out of the chimney. Unfortunately, though British mines were amongst the most efficient in the world at that time, much of this efficiency was the result of mechanisation. Mining the good old fashioned way with pick and shovel may not be much fun for the miners, but it produces exactly the kind of coal you need to run a steam loco on. Mechanized coal extraction produces very small coal, with a high proportion of dust and crumbs. This is perfect stuff for power stations but useless for steam locos – and useless for household heating too. Coal fires were still the most common form of domestic heating at that time. Nobody wants to put small coal on their fires: it's messy, and it tends to get wafted away up the chimney. With increasing mechanisation there wasn't enough large coal to

"Paradoxically, Diesel invented this great engine that made life easier for the six guys who were left in work, and a good deal harder for the hundred who had been laid off."

"With running costs at 50 per cent and a power/ efficiency ratio twice that of steam engines, the big new diesels revolutionized the old railway order."

satisfy both the domestic market and the railway industry. One or the other had to give, and because of its staffing problems, as well as difficulties caused by the Clean Air Acts, the railways decided to convert to diesel.

So did Rudolf Diesel succeed in doing what he set out to do? Is the diesel engine the rational heat engine he envisaged? Well, he certainly didn't abolish the steam engine. The most efficient way of converting heat into motion today is the steam turbine. The small manufacturers whom Diesel wanted to help by creating a small, economical power unit are much more likely to run their machinery off mains electricity, most of which is generated by steam turbines. That said, steam turbines are only efficient when a great power output is required and when that requirement is fairly constant. This is because they take a long time to get going and are very inefficient while they're being started up and shut down. So where smaller-scale or more flexible electricity generation is concerned, the diesel engine is ideal. It is probably true to say that the diesel engine is the prime source of small-scale power to people all around the world who don't have the benefit of mains electricity.

And the marginal benefits of economy and reliability offered by the diesel engine have made it the number one prime mover in all industrial countries. If the petrol engine was to succumb to some unheard-of mechanical virus you might have to get the bus or the train to work instead of driving, but you'd get there all the same – and when you got there, it's more than likely that whatever you happen to do for a living, everything would be going on pretty much as usual. But if there was a virus that attacked only diesel engines, the situation would be rather different. There'd be no trains, no buses and no cabs. There'd be no construction work, no sea freight and no sea transport. There'd be no heat or power in hospitals, where massive diesel-powered-combined-heat-and-power units generate electricity and heat the wards with their cooling water. Agricultural work would come to a complete halt. In short, the diesel engine has insinuated itself into every corner of the industrial world, and that world would simply grind to a halt without it.

CHAPTER FIVE

THE SUPERCHARGER

Tim Birkin

Legend has it that in 1928 Henry R. S. Birkin – better known as Tim Birkin, man of leisure, racing driver and Bentley enthusiast – refused to race his Bentley again until it was supercharged. The Mercedes SSKs which he regularly met on the race track were proving difficult to beat, and Birkin had identified their superchargers as the source of their extra speed. W. O. Bentley himself was opposed to such new-fangled devices. His thoughts on the matter were somewhat bizarre: he described the supercharger as 'a short-cut to higher performance' – as if that were a bad thing in a racing car. He preferred instead to increase horsepower by increasing the size of the engine. Nothing very clever about that.

Fortunately for Birkin his pockets were almost as deep as his convictions, and with a little help from a wealthy female admirer he set up a company to develop and build superchargers. Eventually, Bentley produced a supercharged 4.5-litre model. But the supercharged Bentley was underdeveloped and had reliability problems. Whereas the supercharger on the SSK only came into action when the driver put his foot right to the floor, the Bentley supercharger was on all the time. The strain caused by relentless supercharged combustion in an engine that was not designed to get an entire bellyful of petrol on every intake stroke was too much, and the supercharged cars often failed to finish.

But in theory Birkin was right: no other single modification adds anything like as much clout to an engine's performance as a supercharger. And there's nothing wrong with running a supercharger full-time, so long as the engine is built to take the increased power output. In engineering terms increasing engine size is very definitely the easy way out and supercharging is the high-efficiency alternative.

Roots Type BLOWER

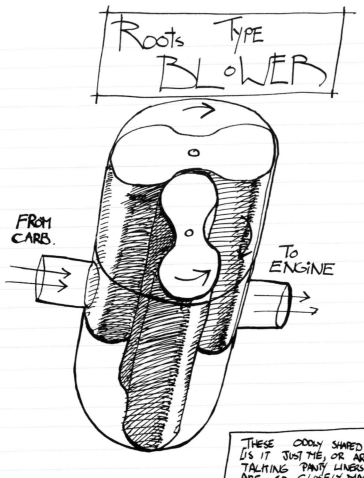

FROM CARB.

TO ENGINE

[THESE ODDLY SHAPED "LOBES" [IS IT JUST ME, OR ARE WE TALKING PANTY LINERS HERE?] ARE SO CLOSELY MACHINED THAT AIR DOES NOT ESCAPE BETWEEN THEM AS THEY ROTATE. THEY PRODUCE A POWERFUL STEADY FLOW OF AIR + FUEL THAT CAN IMPROVE PERFORMANCE BY 40%

END SECTION

Superchargers are compressors. Any kind of compressor will do. One of the best for low-pressure, low-volume work – and the most common type for automobile superchargers in Europe up until the Second World War – was the Roots type. The Roots blower was invented by the American team of Francis and Philander Roots, in 1854. The Roots brothers ran a water-powered textile mill and were looking for an efficient water turbine to provide power for their factory. Francis came up with a twin impeller device. He made it out of wood. The wood swelled in the water and the impellers jammed. But, the story goes, the local foundryman took one look at it and decided it would be ideal as an air-blast blower for his foundry.

The Roots blower (left) has two rotors which mesh together so closely that no air can pass between them. They rotate in opposite directions inside a cylindrical housing, moving a fixed volume of air on each rotation. The Roots design has been used in countless industrial pumping applications, including pumping air into mines. The oil pump in most cars functions on the same principle. When used as a supercharger the Roots blower is always placed between the carburettor and the intake manifold, so it has the added advantage of 'kneading' the fuel and air together, so giving a more even mixture which burns more evenly and more completely.

The supercharger isn't an engine in its own right, it's a type of modification that can be made to reciprocating engines of all sorts. And on the beautiful Bentley and Mercedes sports cars of the twenties and thirties, the supercharger was little more than a bolt-on go-faster extra – albeit a particularly effective one. But if you're not interested in such boy's-own trivia, take heart, because the supercharger had a second, far more important role to play in the history of engineering: it was used to transform the petrol engine from an excellent sea-level power-plant into a half-way decent power-plant for aeroplanes.

Aero engines have to work high up in the atmosphere where air pressure is as little as a quarter of its value at sea level. Without the supercharger, reciprocating engines just can't do the job. And so, without the supercharger there would have been no developmental route between the flimsy canvas, wood and wire contraptions that took to the air before the First World War and the huge passenger planes that – once the jet engine had been invented – brought about what was probably the greatest ever transformation in human transportation.

RIVALRY ON THE RACETRACK — MERCEDES VS BENTLEY

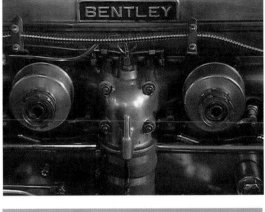

"Seventy years ago, pumping more air into an engine became first a matter of national pride, and then a matter of life and you-know-what."

"The two countries whose national pride were at stake back in those halcyon days were Britain and, yes, you guessed it . . . the Germans."

RIVALRY ON THE RACETRACK — MERCEDES VS BENTLEY

"In Germany, the government were offering very large sums of money indeed to successful grand prix racing teams. I mean, its hardly cricket . . .

. . . in Britain, if you improved the performance of your engine above a certain amount, the government put up your road tax."

"Tim Birkin didn't know what had hit him . . . but as soon as he found out he swore he'd never race again without one."

They don't make them like this any more. A 4.5-litre supercharged Bentley.

And they don't make them like this any more either. A supercharged Mercedes SSK at the Nürburgring in Germany.

"Basically, the supercharger's just a very posh set of bellows. It's suck, suck; blow, blow; whizz, whizz – Wow!"

The supercharger has its effect on the pumping side of a piston engine's business. A piston in a four-stroke engine spends two of the four strokes in its cycle pumping. It pumps exhaust fumes out of the cylinders and fresh fuel/air mixture in. And though the four-stroke does this job far more effectively than the two-stroke, it doesn't do it perfectly. For a start, it doesn't scavenge burnt gases from the cylinder completely. This is because the piston doesn't move all the way to the top of the cylinder, but instead leaves a small part free for the compressed fuel/air mixture to be burned in. This part of the cylinder is called the combustion chamber, and the ratio of the volume of the whole cylinder to the volume of the combustion chamber is the engine's compression ratio. Modern domestic engines have compression ratios somewhere around eight to one. This means that the combustion chamber takes up one-eighth of the volume of the whole cylinder. It also means that at the end of the exhaust stroke approximately one-eighth of the total volume of the cylinder will still contain exhaust gas. This exhaust gas remains in the cylinder when the fresh fuel/air is sucked in. It reduces the total amount of fresh fuel/air mixture that can be drawn into the cylinder – by around one-eighth – and at the same time has an inhibiting effect on the combustion. And on top of this the fuel/air intake itself is far from perfect. When the intake valve opens and the piston starts to move back down the cylinder it leaves the space which it just occupied empty. This causes a partial vacuum, into which atmospheric pressure pushes fuel/air mixture from the carburettor. But this has to happen very quickly – 50 times a second in an engine that is revving at 6000 r.p.m. – and the fuel/air mixture has to pass through pipes and valves which impede its flow. In consequence the cylinder gets much less than the theoretical maximum charge it is capable of receiving.

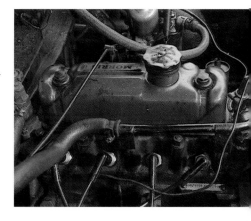

"You need three things to make an explosion: fuel, heat and air. And the more fresh air you can get in to the equation, the bigger the explosion gets."

These problems are expressed by engineers in terms of an engine's volumetric efficiency. A cylinder that has a one-litre capacity which is successfully charged with one litre of fuel/air mixture has a volumetric efficiency of 100 per cent. But because of the inefficiencies mentioned above, the volumetric efficiency of working engines tends to be considerably lower than 100 per cent. Many of the go-faster 'tuned' bits that people who are that way inclined bolt on to their engines to replace the less efficient bits that come from the factory are specifically

INTERCOOLERS

A veteran of the Le Mans 24-hour race once told me that his car went 20 m.p.h. faster down the Mulsanne straight at night than during the day. The decrease in air temperature caused a kind of natural supercharging. As air cools down, its molecules become less energetic and so its density increases. This means that the same cylinder volume will contain a little bit more air, and a little bit more fuel to go with it. So the quantity of charge that is compressed and ignited in the cylinder goes up. Turbocharged and supercharged engines with intercoolers intentionally create the same effect. Cooling the pressurised fuel/air mixture increases its density and so increases the mass of fuel/air mixture that can be crammed into the cylinder. Intercooling has the added advantage of allowing the charge to be compressed further before it reaches combustion temperature. It is possible to 'supercharge' an engine by using an intercooler on its own.

designed to improve volumetric efficiency. Twin intake and exhaust valves, overhead camshafts with high-lift cams, wider and smoother intake and exhaust manifolds, and double exhaust pipes are all intended to improve the flow of gases into and out of the cylinder.

But instead of messing around making small improvements to the flow of gases, you can revolutionise the whole process by adding a supercharger. A supercharger is an air pump that takes its power

COMPRESSION RATIOS

The limit for a petrol engine's compression ratio is a function of the combustion temperature of its fuel. Compression ratios in domestic vehicles have increased over the years from 3 to 1 in very early petrol engines to 8 or 9 to 1 in domestic cars and as much as 12 to 1 in race cars. In line with Carnot's theories (*Diesel*), this increase in compression ratios has yielded a massive improvement in engine power and efficiency. Apart from improvements in casting and machining techniques, the possibility of super-efficient high-compression engines is largely the result of the improvement in petrol technology. It's all to do with octane ratings. A fuel's octane rating is a measure of its ability to resist pre-ignition (sometimes known as knocking) due to the adiabatic increase in temperature during the compression stroke. Crude oil that comes out of the ground is distilled into fractions of differing volatility – propane and butane being the most volatile, then petrol, then paraffin, diesel, fuel oil, lubricating oil and finally bitumen. The petrol fraction is further refined before it becomes suitable for running car engines on. The refining process alters the chemical composition of the petrol in order to increase its resistance to pre-ignition. Additives such as lead have the same effect.

In case you ever wondered, a fuel's octane rating is equal to the proportion of isoctane in a mixture of isoctane and heptane which has the same anti-knock performance. You can't have a mixture that contains more than 100 per cent isoctane, so for ratings higher than 100 the octane rating refers to the millilitres of tetraethyl lead that has to be added to a gallon of isoctane to get an equivalent anti-knock performance.

A supercharger boosts the pressure in the engine's intake manifold. This has the effect of increasing the engine's overall compression ratio. So in order to ensure that the fuel doesn't reach combustion temperature and ignite before it is meant to, the compression ratio of a supercharged engine must be reduced considerably. This doesn't result in a loss in efficiency because the mixture is being compressed just as much in the end – the difference being that in a supercharged engine the work of compression is shared between the supercharger and the piston.

TURBOCHARGERS

In recent years supercharging has been replaced by turbocharging as the most popular way to improve the volumetric efficiency of car engines. A turbocharger differs from a supercharger only in the way it is powered. Whereas superchargers are driven off the crankshaft, turbochargers are driven by a turbine – a kind of hot-air windmill – placed in the path of the exhaust gas.

Invented by the Swiss engineer Büchi in 1908, the turbocharger was intended as an improvement to the diesel engine. Despite its origins, turbocharging was favoured more outside Europe, while European manufacturers concentrated their attention on developing the supercharger. Developing turbines that could rotate at the terrifically high speeds necessary to drive a compressor while at the same time standing up to the extreme heat of exhaust gases was a tricky business, and it was a long time before they became reliable.

In recent years turbochargers have been favoured over superchargers by car manufacturers. Their supporters argue that turbochargers supply the power to turn the compressor for free, while superchargers absorb a considerable amount of the crankshaft's power output. It's true that superchargers absorb a lot of power, but it isn't true that a turbocharger gets its power for free. The turbocharger's turbine creates considerable 'back pressure' in the exhaust system and inhibits the flow of exhaust gases from the cylinders. So although the net effect is still positive, to some extent the turbo takes away with one hand what it gives with the other. And though it is technologically quite possible these days, making reliable turbines is still a very expensive business. So superchargers have some important advantages over turbochargers.

from the crankshaft. It can be located before or after the carburettor has mixed petrol with the intake air. It improves an engine's volumetric efficiency by increasing the pressure in the intake manifold – and therefore the density of the fuel/air mixture entering the cylinder. This has several effects, all of which are good. Firstly, the increased intake pressure means that when the intake valve opens and the piston starts moving down the cylinder the mixture is pushed into the cylinder with greater force. The net effect of friction and turbulence is reduced and so more mixture finds its way into the cylinder. Secondly, because the pressurised mixture is more dense than it would be at atmospheric pressure, a given volume of the pressurised mixture contains more fuel and more air than the same volume of unpressurised mixture. So it is possible to cram more fuel/air mixture into the cylinder than even 100 per cent volumetric efficiency would allow. This means a supercharged 2-litre engine may be made to burn as much fuel as, say, an unsupercharged 2.5-litre engine – while retaining the advantages of smaller size and weight. The third advantage is that because the fuel/air mixture enters the cylinder under higher than atmospheric pressure, during the brief 'overlap' period when the intake and exhaust valves are both open, the fuel/air mixture can have the positive effect of pushing some of the remaining exhaust gases and out of the cylinder, making room for more fuel/air mixture in the cylinder and giving a more efficient combustion.

As well as the traditional advantages of a supercharged engine, this Mercedes Benz SLK 230 Kompressor also uses compressed air from its supercharger to clean the exhaust when starting from cold. Air injected into the exhaust helps oxidise uncombusted components in the exhaust gas, raising exhaust temperature and causing the catalytic converter to heat up more quickly.

Roughly speaking, that's the story as far as supercharging cars is concerned. It's a neat trick but it could hardly be described as world-changing. But as soon as you put a piston engine in an aeroplane – and until the jet engine was invented in the late thirties you didn't have any choice in this respect – the supercharger becomes very necessary indeed. The reason is the rapid fall in air pressure that occurs as you travel up through the atmosphere. This thinning of the air in the inlet manifold has precisely the opposite effect to supercharging. It makes the fuel/air mixture more sluggish going into the cylinder, reduces the density of the mixture in the cylinder and so reduces the mass of fuel that can be burned. So as a piston-engined aeroplane goes up in the atmosphere the power output of its engine goes down.

This is a serious enough problem in itself, but when you take

"I suppose the super-charger really belongs to that bygone age when the measure of a man's worth was his capacity for devil-may-care recklessness or unselfish heroism."

into account how propellers work it becomes more critical still. A propeller is a rotating aerofoil. It works just like an aeroplane's wing, creating a pressure difference between the air that flows on one side of it and the air that flows on the other. The net effect of this difference is that the air on the high-pressure side of the flow exerts a force on the wing or propeller. In the case of a wing this force is called lift; in the case of the propeller it is called thrust. The reason why a propeller doesn't look exactly like a wing is because the speed with which a propeller moves through the air varies along its length. For an aerofoil to work with maximum efficiency its angle of attack must be altered to match its speed. So since the tip of a propeller blade moves much faster than its root, propellers have a twist in them. The twist is designed so that the propeller's angle of attack decreases from root to tip.

Incidentally, the usual 'simple' explanation for why aeroplane wings fly is a bit misleading. The explanation centres around the fact that the wing has an asymmetrical profile which forces air to travel further around the top of the wing than around the bottom of the wing, obliging it to speed up and lose pressure as it does so. Anyone who's seen an aeroplane fly upside down will have had their doubts about this. If wings fly quite well upside down – and they obviously do – then it can't really be their shape that's most important.

It isn't. Early biplanes had wings which were almost flat. The fat 'aerofoil' shape was developed because the wires and struts necessary to keep flat wings stiff and strong were not possible with a monoplane – and anyway, they caused an unacceptable amount of drag. Fat wings can be strengthened from within. It was a lucky break that fat wings have some aerodynamic advantages too.

What is essential if lift is to be generated is that the wing should have a positive angle of attack. This means that it should be tilted upwards to some degree. The faster the airflow over the wing the smaller the angle of attack has to be to generate lift. The asymmetrical aerofoil-type wing has its positive angle of attack built into its shape and has to be actually tilted downwards if it is to generate no lift at all.

At low speeds the airflow over a wing is symmetrical and although there are speed and pressure differences within the flow on

"But then sixty years later we find a production car that has as standard a supercharger – and I wonder if you can guess who made it?"

CARBURETTORS AND THE VENTURI EFFECT

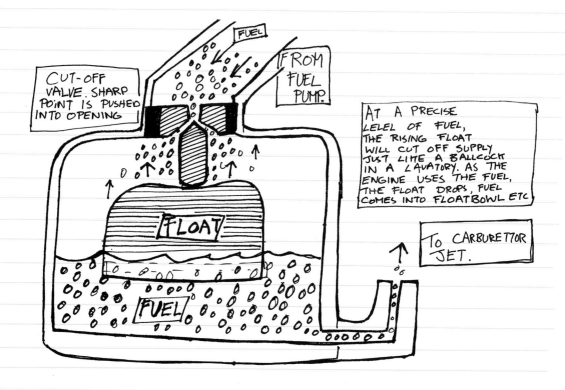

CUT-OFF VALVE. SHARP POINT IS PUSHED INTO OPENING

FUEL

FROM FUEL PUMP.

FLOAT

FUEL

AT A PRECISE LELEL OF FUEL, THE RISING FLOAT WILL CUT OFF SUPPLY JUST LIME A BALLCOCK IN A LAVATORY. AS THE ENGINE USES THE FUEL, THE FLOAT DROPS, FUEL COMES INTO FLOATBOWL ETC

TO CARBURETTOR JET.

Carburettors work by exploiting the venturi effect to suck petrol vapour into the air flow. The venturi effect gets its name from the eighteenth-century Italian mathematician and physicist, Giovani Battista Venturi. Venturi studied the behaviour of fluid flows. What he discovered flies in the face of common sense. When air or any other fluid is flowing through a constriction in a pipe you would expect the fluid to slow down and pressure to build up as its molecules 'queue' to pass through the narrow gap. But what actually happens is the exact opposite of this: the air speeds up and its pressure falls.

One way of understanding why this should be so is to think in terms of flow rates. The air flows at a rate of a given mass per second. When the air reaches the constriction the speed of the flow has to increase in order to maintain the same mass flow rate.

The reason why an increase in flow velocity results in a decrease in pressure was discovered by the Swiss scientist Daniel Bernoulli (1700-1782). The pressure that a gas exerts in all directions simultaneously is known as its static pressure. But gases that are flowing also exert dynamic pressure. Dynamic pressure is the pressure that blows your hair back when you're riding in an open-top car or a motorbike, and the pressure that the wind is exerting when it rustles the leaves on the

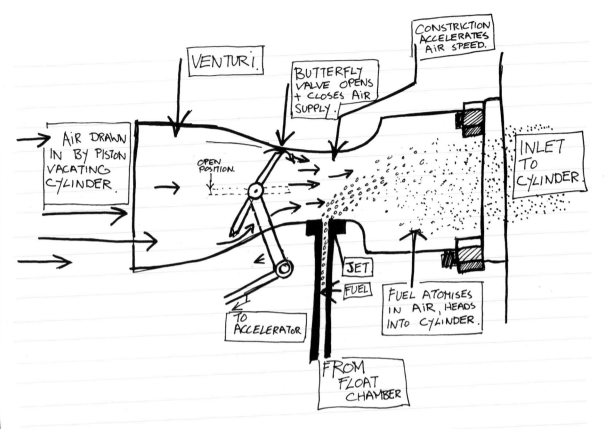

- **VENTURI.**
- **BUTTERFLY VALVE OPENS + CLOSES AIR SUPPLY.**
- **CONSTRICTION ACCELERATES AIR SPEED.**
- **AIR DRAWN IN BY PISTON VACATING CYLINDER.**
- **INLET TO CYLINDER.**
- OPEN POSITION.
- **TO ACCELERATOR**
- **JET FUEL**
- **FUEL ATOMISES IN AIR, HEADS INTO CYLINDER.**
- **FROM FLOAT CHAMBER**

trees. Bernoulli discovered that so long as no energy is added to or removed from the gas, as dynamic pressure increases static pressure decreases, and as dynamic pressure decreases static pressure increases. This is because at the molecular level the pressure exerted by a gas on any object is the sum total of the tiny forces produced when the gas molecules go crashing into the object. The total kinetic energy contained in the gas is expressed by its static pressure plus its dynamic pressure. As long as no energy is added to or taken away from the gas, the energy it contains will remain constant, and so the sum total of the two kinds of energy will remain constant also.

NICE PONG.

SCENT

The Venturi principle has lots of other applications, too . .

each side of the wing, there is no net pressure difference between the upper and lower surfaces. But when a certain speed is reached the flow becomes asymmetrical. A net speed and pressure difference results and lift is generated.

The reason why the flow takes the shape it does is horribly complex. The usual explanation – that the air has to travel faster over the top of the wing in order to 'catch up' with the flow under the wing – just doesn't get close. The full explanation has to do with the viscosity of air. We don't tend to think of air as being sticky but there you go. When a wing moves through the air a tiny layer of the air behaves as if it is 'sticking' to the wing and is dragged along. At low speeds this viscosity results in a symmetrical flow, while at higher speeds the stickiness 'fails', the pattern changes and an asymmetric flow results.

When it comes to flying at altitude the fact that both aircraft wings and propellers work as aerofoils leads to a vicious circle. The amount of lift developed by a given wing is dependent upon its speed through the air and the density of that air. In the lower air densities found higher up in the atmosphere the lift a wing produces at a given speed falls. That's not a problem in itself; all you have to do is to fly the plane a bit faster. To fly the plane faster, of course, you need a little more thrust from your propeller. No problem in itself; you only have to spin the propeller round a bit faster. Which wouldn't cause any difficulty at ground level, where there's plenty of oxygen around, but in the upper air, where the air is thinner, engine output has started to drop . . .

One solution would be always to fly close to the ground. But the advantages of flight in the upper part of the atmosphere are considerable. Bomber and fighter planes both reap tremendous benefits from being able to fly high up in the atmosphere, and apart from the tactical military advantages, because the air is thinner the higher up you go the drag that impedes the forward movement of aeroplanes through the air is greatly reduced. In level flight at a fixed speed the power needed to propel a plane forward is exactly equal to the drag acting on the plane. So if drag is reduced by half, so is the plane's fuel consumption.

THE AEROFOIL
THE SIMPLE VERSION.

LIFT

AEROFOIL PROFILE

AS THE WING PASSES THROUGH THE AIR, THE AIR ON THE FLAT BOTTOM TRAVELS A MUCH SHORTER DISTANCE IN THE SAME TIME AS THE AIR ON THE TOP SURFACE, WHICH IS OBLIGED TO TRAVEL FASTER. THE HALF-VENTURI SHAPE ON THE TOP ALSO HELPS TO ACCELERATE THE AIR TOWARDS THE BACK. THIS HIGHER SPEED PRODUCES A HUGE PRESSURE DIFFERENTIAL, RESULTING IN LIFT.

RIVALRY IN THE AIR – MESSERSCHMITT V SPITFIRE

"With its supercharged Daimler-Benz engine producing 2,000 horsepower and flown by ace pilots, early encounters with the Messerschmitt ME 109 were often literally devastating to the RAF."

"There's a world of difference between flying as fast as you can in a figure eight and engaging in an aerial dogfight."

"The Spitfire was far easier to handle - but in many ways the Messerschmitt was faster . . ."

RIVALRY IN THE AIR — MESSERSCHMITT V SPITFIRE

"I can't make my mind up - should I opt for the British model with the go-faster stripes . . .

. . . or the Continental model with the beige finish?"

"All these fantastic technical advances came about so that we could kill each other more efficiently - that's what depresses me."

CENTRIFUGAL COMPRESSOR

After initial experiments with 'positive displacement' air pumps of the Roots type – and, in the case of the RAF, a brief flirtation with the simple piston and cylinder version (which works just like a bicycle pump) – the centrifugal type became the compressor of choice for aeronautical engineers struggling to push their planes higher and faster.

The centrifugal compressor is another device that exploits Bernoulli's principle. Air is sucked in at the centre of the rotating vane wheel and accelerated outwards by centrifugal force. This adds kinetic energy to the air. When the air leaves the vane wheel it is slowed down again.

The kinetic energy that has been added doesn't go away, but is converted into static pressure energy. In this way the centrifugal compressor increases the pressure and density of the gas that passes through it.

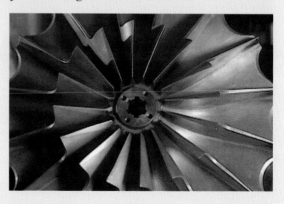

That's why the supercharger quickly became essential equipment on any plane that had ambitions to stray into the upper air, and why the development of the supercharger was pushed on with so swiftly during the Second World War. The battle for air superiority over Europe was very much the battle to force the development of the supercharger. The effects of this development on the performances of the high-altitude fighters was staggering. The top- performing British fighter, the Supermarine Spitfire, was powered by the supercharged 27-litre V12 Rolls Royce Merlin. When it first flew in March 1936 the Spitfire was capable of 355 m.p.h. at 19,000 feet. The first versions of the Merlin were fitted with a single-speed, single- stage supercharger and developed around 1000 hp. The course of development took the Merlin through some thousand or so modifications. As the single-speed, single-stage supercharger evolved into a two-speed, two stage compressor with intercooler, the engine's power output, rated at just over 1000 hp at the beginning of the war, had by 1944 increased to over 2000 horsepower. This increase in power – and, more importantly, in high-altitude power – took the

GYROSCOPIC PRECESSION

If you've ever played with a gyroscope you'll probably remember the bizarre habit they have of shooting off in completely the wrong direction when pushed. This effect is called gyroscopic precession. A force applied perpendicularly to the axis of rotation of a gyroscope is rotated 90 degrees in the direction of spin, resulting in a force acting perpendicularly both to the axis of rotation and to the first force. So if you push a clockwise rotating gyroscope north it moves east.

A quickly rotating propeller has a considerable gyroscopic effect. So when a pilot lifts the tail of an aircraft off the ground in preparation for take-off, there is a gyroscopic force of precession to the left or the right, depending on the direction in which the propeller is turning.

On top of this force there's the torque reaction from the engine. While the engine is applying force to the propeller to turn it in one direction, there is an equal and opposite turning force exerted on the engine – and therefore the plane – in the other direction. This means that one wheel digs into the tarmac harder than the other – causing a turning effect in the same direction as the gyroscopic precession.

And on top of these two factors there's the spiral slipstream created by the propeller, which hits the plane's vertical stabiliser in such a way as to create a turning effect – in the same direction as the gyroscopic precession. The more powerful the engine the stronger these effects become. The pilots of the most powerful fighters at the end of the war needed all their strength just to keep their planes running straight down the runway for take-off.

"At the end of the day, it's just a bloody great hair dryer."

Spitfire to a maximum speed of over 400 m.p.h. at just under 28,000 feet, and would allow it to go on climbing – albeit at reduced speed – way beyond that height.

The supercharger took the reciprocating aero engine right to the limit of its performance. But it really was the limit. The fighter planes at the end of the Second World War put out so much exhaust that pilots had to put on their oxygen masks before they started up their engines to avoid carbon monoxide poisoning. The handling of planes with such massively powerful engines was problematic: one World War Two pilot described taking off for the first time in a Typhoon and, not being fully prepared for the immense gyroscopic forces created by such a powerful engine, finding himself leaving the ground between two hangars – at right angles to the runway. The noise and vibration created by these monstrous engines was fearsome. This wasn't an important consideration for military purposes, but was a very real drawback when it came to persuading people to abandon the luxurious ocean-liners that carried the majority of the world's long-distance passenger traffic. On top of which the fuel economy of piston aero engines wasn't great. Travelling any distance by air involved many refuelling stops, and this was reflected in the price of air tickets. So although big transport planes had developed all the essential features of the modern airliner, commercial air travel was a transport revolution in waiting.

What it was waiting for was the engine that could power planes safely, cheaply, and comfortably across long distances – in a word, the jet. The jet is so much more efficient, so much more powerful, so much quieter and more reliable than the best piston aero engines, that – although there's still a lot of life left in the supercharger for cars and trucks and boats and trains and all things that creepeth upon the earth – as far as aero engines are concerned it has been well and truly consigned to history. But as we'll see in the next chapter, the supercharger was more than just an interim solution: without the development of the centrifugal compressor for use in superchargers, the jet engine would never have been possible at all.

CHAPTER SIX

JET

At the naval review at Spithead in 1897 the most advanced triple-expansion marine steam engines were rendered old-fashioned on the spot when the Turbinia came skimming down the Solent at her maximum speed of 34 knots. The Turbinia was the first ship to be fitted with one of Charles Parsons' steam turbines. It was both faster and more economical than existing reciprocating engines. What has this to do with the jet engine, you may well ask? Bear with me and all will become clear.

Charles Parsons specialised in high-speed steam engines. He built his first steam turbine in 1884 when the company he was working for asked him for a high-speed steam engine to power a dynamo. It was a success, producing 10 hp at 18,000 r.p.m. Over 300 were sold in the next five years for generating electricity on board ships. Parsons went for a turbine because accelerating and decelerating the pistons of a reciprocating engine hundreds of times a second produces massive stresses. And because they produce rotary motion indirectly via a crankshaft, reciprocating engines generate an awful lot of friction. Turbines produce rotary motion directly, with a minimum of stress and a minimum of friction, so they can turn much faster than reciprocating engines and are mechanically more efficient. Steam turbines also have the advantage of being able to extract work from extremely low-pressure steam – steam that is utterly useless to a reciprocating engine.

In 1889 Parsons set up his own company and in 1894, using his own money, he built the Turbinia in order to convince the Admiralty that the steam turbine was the next big step in steam technology. The Admiralty wouldn't be persuaded, though, so Parsons staged a little demonstration for their benefit. In 1897, the year of Queen Victoria's Jubilee, the Royal Navy had organised a massive celebratory review of both British and foreign vessels. Parsons came steaming down from

"The engine that's probably had most influence on the world this century, the most simple, the most powerful, most reliable engine in the history of engineering, is very much the Cinderella of the piece."

the Tyne and gatecrashed the review, whizzing up and down the lines of ships and generally showing off in the most un-English fashion imaginable. The Navy sent their fastest ships out to put a stop to Turbinia's antics but they couldn't get anywhere near her. The Admiralty were humbled. Forced to recognise the advantages of Parsons'engine they commissioned a turbine-powered destroyer, HMS Viper. They still required that Parsons should deposit £100,000 as security however – just in case the new vessel wasn't all that Parsons cracked it up to be.

It was the turbine that took steam beyond its apparent limits and opened up a whole new level of performance, taking it beyond the realms of the triple expansion engine into a land of undreamed-of efficiencies. Years later the turbine's advantages over the piston asserted themselves again, this time in the field of the internal combustion engine. And eventually, the turbo-jet would render reciprocating aero engines defunct in just the same way. (The inventor of the turbo-jet even had a similar struggle to persuade the armed forces of his invention's usefulness.) And although the death of the piston engine was never marked in such fine style as Turbinia's dramatic display, the engine that replaced it in the air brought about the most spectacular transport revolution in human history.

The limitations of the piston engine and propeller combination had been forseen more than twenty years before the outbreak of the Second World War by a young RAF officer cadet by the name of Frank Whittle. In his final year essay, 'Future Developments in Aircraft Design', Whittle calculated that a wind of 100 m.p.h. against a plane

"If you think the silicon chip has revolutionized the twentieth century, think again. Surfing the net has nothing on a fighter plane doing Mach 2 at sea level, or a fully-laden 747 doing 500 m.p.h. at 38,000 feet."

going at 600 m.p.h. at a height of 120,000 feet would cause less drag than a 20 m.p.h. wind acting against the same plane travelling at 1000 feet. He predicted that the future of flight would therefore be at high speeds in the upper atmosphere, where air resistance to the forward motion of the plane is drastically diminished, fuel efficiency is far greater, and air turbulence is reduced. Whittle was contemplating speeds of 500 m.p.h. at 60,000 feet, at a time when the air speed record stood below 300 m.p.h., the fastest RAF planes could not top 150 m.p.h. and the maximum height they could attain was less than 15,000 feet.

WHITTLE'S EARLY LIFE

Frank Whittle was born in Coventry on 1 June 1907. His parents, both strict Wesleyans, were uneducated Lancashire mill workers, but his father, Moses, was ambitious, and while working as a machine-shop foreman with a firm in Coventry, saved up enough money to buy the Leamington Valve and Piston Ring Company, a one-man engineering firm in Leamington Spa. Moses was an ingenious man but, having been forced into the mill at the age of ten, had insufficient education to make best use of his talents. He devoted much of his time to that old engineers pipe-dream, the perpetual motion machine. His business did well enough out of the First World War to enable the Whittles to buy a car, but after the war business took a downturn and the family was thrown out of their home. During Frank Whittle's youth poverty was never far away.

As a boy he was bright at school, and won a scholarship to Leamington College. He went on to win a scholarship to grammar school, but his family could not afford to let him take it up. He cashed in the scholarship and continued at Leamington. But being independently minded from an early age and finding reason to dislike almost any activity in which he was obliged to take part, he preferred to follow his own interests at the local library (sometimes playing truant in order to go there) where he studied engineering, shorthand, physiology and comparative religion – and anything else that took his fancy. He studied aircraft engineering so vigorously that he was convinced that, given the chance, he would be able to fly a plane without instruction. He was good at daredevil stunts but loathed games, despite being an able athlete, preferring to stay in the school chemistry lab performing experiments and – just as you'd wish in a boy his age – manufacturing explosives. The only sport he showed any enthusiasm for was the marathon (which was optional), in which he came fourth.

Whittle was well aware of the problems thin air causes for piston-driven propeller planes (see *Supercharger*), so in his essay he explored the possibilities of another form of propulsion that would enable aeroplanes to reach their full potential. What this form of propulsion was to be, the young Whittle couldn't decide. He worked it out soon enough, however: it was the jet engine, and within eight years of writing his essay Whittle had designed and built one. A few years after that he had a jet-powered aeroplane in the air.

Anyone who holds to the silly and envious theory that great inventors are motivated by the need to compensate for perceived personal inadequacies will find plenty of grist for their mill in the life of Frank Whittle, which was not short on petty humiliations. Whittle's most striking physical characteristic was his size. He was a very small man indeed. During his first few weeks at the RAF college at Cranwell he suffered the indignity of going on parade in a lounge suit and a beaten-up old bowler hat, because the RAF didn't have a uniform small enough for him and had to have one specially made.

From a working-class background, he became an RAF officer the hard way, starting out as a technical apprentice. But his size almost prevented him from making it into the RAF at all. He first applied in 1922, at the age of fifteen, passing the written exam without difficulty, but was turned down on the grounds of his height and chest measurement. He was only five feet tall at the time. A sympathetic RAF physical training instructor came to his rescue, recommending a nourishing diet – including a daily dose of olive oil – and a list of exercises from the 'Maxalding' system. The system must have been a good one, because the stunted young man grew three inches in six months, and expanded his chest by the same amount. Even so, he was rejected a second time, and strictly speaking, that should have been that. But Whittle applied again, pretending not to have been through the selection procedure before. This time he was successful.

At the end of his apprenticeship Whittle came sixth out of six hundred. Only five places as Flight Cadets were to be given to ex-apprentices, but the boy who came top failed the medical and Whittle squeezed through. At RAF Cranwell he was a fish out of water: a

"Back in the 1920s, getting around was all about oily smells, filthy hands, and plotting your trip from one garage to the next."

THE AVRO 504K

Frank Whittle learned to fly in a castor-oil-lubricated, fabric, wood and wire 1918 Avro 504K – receiving only five and a half hours of dual-control instruction before going it alone. The Avro 504k was powered by a rotary engine. In most engines the cylinders are fixed and the crankshaft turns: in the rotary engine it works the other way around. The crankshaft is fixed to the body of the plane and the body of the engine is allowed to rotate around it. The propeller is fixed directly to the engine. This system saves some weight; if you hold the engine stationary it is necessary to provide a flywheel to help smooth over the essentially lumpy four-stroke power production process, absorbing some of the energy produced by combustion strokes and returning it when no power is being produced. The rotary engine dispenses with the necessity of a separate flywheel. By rotating around a fixed crankshaft the engine acts as its own flywheel.

working-class apprentice surrounded by public school boys. Nocturnal beatings were rife – as were all the usual displays of that peculiar version of culture and breeding so favoured by English public schools. Whittle submitted to such treatment without complaint. He was determined to become a member of the club.

His strategy was to win acceptance by excelling. His academic performance was brilliant; and he did well as a pilot too. It soon became evident that he was a bit of a daredevil – not quite what you'd expect from one of the greatest engineering minds of the century. He wrote off his first plane after only a few hours as a solo pilot. Poor visibility had forced him to land in a farmer's field in order to ask where he was. Having ascertained his location he then went against regulations by taking off again. The field was soft, making his take-off slow, and he forgot to take into account the fact that there was a cross-wind, which took him straight into a tree. Whittle himself was unhurt.

A little later in his training Whittle got in trouble with the police for 'hedge-hopping' in a Bristol fighter. Soon after that he was disqualified from the end-of-year aerobatics competition on the grounds that his tricks were too risky. But his most spectacular display of airborne carelessness must have been while he was working as a flying instructor. While practising 'crazy flying' for an air show Whittle infuriated his flight commander by writing off two planes in only three days. He was something of a liability on a motorcycle too, crashing on many occasions – once going head-on into an omnibus. He never seems to have suffered much in the way of injuries, though he was frequently in trouble with his superiors. In general his exemplary academic record and high all-round ability got him off the hook. Only on one occasion did he have to resort to desperate tactics, making a personal appeal to the family of a pregnant woman who had been disturbed by his unauthorised aerobatics over Canvey Island. Facing the possibility of a court martial and probable discharge from the RAF, Whittle successfully persuaded them not to press charges.

However erratic they may seem in retrospect, Whittle's flying skills must have impressed the RAF, because after spending a short time as a regular fighter pilot he was transferred to a post as a flying instructor. It was in 1929, while he was training to be an instructor, that he came up with his idea for a jet engine. The daredevil strand continued in 1931 when Whittle was sent to RAF Felixstowe for a year to work as a seaplane test pilot. It isn't clear whether this was an expression of the RAF's high opinion of his flying abilities or the their desire to put his taste for danger to better use than terrorising pregnant women. One of the more hair-raising activities he was involved in at Felixstowe was crashing floatplanes into the sea in order to allow their designers to study how well the planes coped. Whittle himself seems to have coped extremely well, taking the excitement in his stride despite the fact that he was a non-swimmer. Like all good test pilots, he was heavily involved in the technical side of things, so the work was ideally suited to his twin passions for danger and engineering. As if this wasn't impressive enough, it was while he was working as a test pilot that his first turbo-jet patent was published.

"The jet has no domestic application. The man in the street has never taken one apart or been in charge of one."

The jet engine is probably the least generally understood of all engines. This might have something to do with the fact that most people who know about engines learn by tinkering with them, and for obvious reasons, tinkering with jet engines is discouraged, by parents and air-safety groups alike. The most we normally get to see of a jet engine is the engine casing and the fan at the front. And the fan isn't even an integral part of the engine. The first jets did without them altogether. But despite this popular obscurity the jet engine is actually the simplest of all internal combustion engines.

Whittle called his invention the Gyrone. The name didn't stick, however; today his invention is known as the turbo-jet. This is usually shortened to 'jet', but the full name is more descriptive, suggesting a combination of two distinct ideas: the gas turbine and jet propulsion. The gas turbine is a particular type of internal combustion engine which can be used for all sorts of purposes other than jet-propulsion, including electricity generation and providing power for high-performance warships, hovercraft and helicopters. Gas turbines small enough to power lorries and cars have been built, but the gas turbine is only fuel-efficient when it is run at optimum speed, so it is not well suited to applications in which engine speed has to vary a great deal.

The big difference between gas turbines and piston engines is that gas turbines run on a cycle of constant combustion. This means that compression, combustion and exhaust take place constantly and simultaneously – which gives the gas turbine a power-to-weight potential greater than any reciprocating engine, and makes it especially suitable as a power source for aeroplanes, whether by means of a propeller or by jet propulsion.

"This, believe it or not, is a primitive turbine..."

The turbine itself is a very ancient idea. A water wheel is a kind of turbine, as is a windmill; and from the middle ages onwards hot-air turbines known as smokejacks were used to turn meat spits in the kitchens of large houses. By the nineteenth century they were standard issue. The smokejack is a kind of gas turbine, but without compression. The turbine is placed in the chimney. The hot air from the fire flows up the chimney, forcing the turbine to rotate, and the rotating motion is transferred to the spits via a system of pulleys or gears.

METALLURGY

The jet engine was a long time being invented – but not because of its complexity. Like the high-pressure steam engine, the jet engine was a well-worn idea waiting for metallurgy to make it possible in practice. High-pressure steam had to wait for steel sufficiently strong to make safe high-pressure boilers, while the jet engine waited on advances in compressor design and the development of metal alloys that would stand up to the high temperatures that turbines have to cope with. Reciprocating engines reach similarly high temperatures but only for tiny fractions of a second – and these hot periods are buffered by relatively long periods during which combustion is not taking place.

"The compressor is basically a huge fan that sucks air into the engine and squeezes it into an explosive mixture with the fuel . . .

. . . This then ignites and goes skooshing out the back, creating thrust, and turning the turbine on its way. Because they're both connected to the same shaft, the turbine then turns the compressor again, so completing the cycle."

THE GAS TURBINE

The gas turbine was patented as early as 1791 by an Englishman called John Barber, although the first gas turbine to be built and run was the Stolze Hot Air Turbine, designed and built by Dr F. Stolze of Berlin between 1872 and 1904. This was an external combustion turbine which used hot air as its working medium. Air was compressed and then heated by being piped through a furnace. It was then allowed to expand through a turbine. The turbine powered the compressor and still had some shaft-power left over to do useful work. The Stolze engine was remarkably modern. Both the compressor and the turbine were of the multi-stage, axial-flow type – just like in a modern jet. This was Stolze's downfall, however, since the axial-flow compressors of the day were hopelessly inefficient. The first working jet engines used centrifugal compressors, and axial-flow compressors were not perfected until the middle of the twentieth century.

Nevertheless, the early years of the twentieth century saw many attempts to build gas turbines, of which the most successful in Europe was the Armengaud-Lemale gas turbine, developed with the help of the Swiss engineering firm Brown Boveri. It used 25 centrifugal compressors in sequence to produce about 300 horsepower, but it was no use in industry, since it consumed four pounds of petrol per horsepower hour, while piston petrol engines of the day consumed an eighth of that. In the States a similar design was developed which showed a similarly astronomical fuel consumption. In both cases the efficiency of the engine was limited by the maximum heat the turbine could withstand. Both firms' work resulted efficient centrifugal compressors, however, which sold well for other purposes.

WHITTLE AND VON OHAIN

Whittle wasn't the only person to invent the jet engine. A young German by the name of von Ohain worked on the idea independently. Whittle's first patent pre-dated von Ohain's work, and Whittle made the first working jet engine – but von Ohain was responsible for the first jet-powered flight by a margin of some two years. The honours for the invention should be shared equally between the two men, but because of the German defeat – and the destruction of the factories that were producing German jet planes – from the point of view of the subsequent history of the jet, von Ohain's work counted for little. The Russians salvaged what they could of the pioneering German work, but in the end, both American and Soviet jets were direct descendents of Whittle's prototypes, and the first jet-powered civilian planes of de Havilland and Boeing have their ancestry in the same lineage. So as a matter of historical influence, Whittle was the pioneer of the jet age even though von Ohain got a jet plane in the air before him.

The surprising thing about gas turbines is that, compared to reciprocating engines – and especially compared to the massively complex reciprocating aero engines they replaced – they are so very, very simple. A basic jet engine has only three main components: a compressor, a combustion chamber and a turbine. The compressor and the turbine are joined together by a shaft and rotate as one. The compressor pressurises atmospheric air and pushes it into the combustion chamber.

Fuel – usually paraffin – is sprayed into the combustion chamber where it vaporizes and burns. The turbine is placed in the way of the expanding gases released from the other end of the combustion chamber. These highly energetic compressed gases expand, accelerating out of the combustion chamber and through the turbine, which is forced to rotate. So as the hot gases turn the turbine the compressor is cramming more air into the combustion chamber where more fuel is being burnt, producing more hot gases to turn the turbine, and so on.

Jet engines needn't be powered by gas turbines. At the most basic level, if you blow up a balloon and let it go you have a fully functioning and completely useless jet-powered aircraft without a gas turbine in

It's easy to imagine that a jet must work by pushing against the air behind the engine. But rockets use jet propulsion, and they keep working out in space where there's no air at all. The difference between a turbo-jet and a rocket lies only in the way in which the expanding gases are generated. Combustion is a chemical reaction between the substance we call the fuel and oxygen. A turbo-jet carries fuel only and takes its oxygen from atmospheric air as it goes. A rocket carries fuel together with oxygen or some oxygen-rich compound, so it doesn't need atmoheric air at all and can go places where turbo-jets can't.

If you're wondering how come the compressed air can drive the compressor and still have energy left over to provide thrust, you only have to remember Carnot's theory (see *Diesel*), which applies equally to turbines as to piston engines. The compressed air has heat added to it before it is allowed to expand, so it is able to do more work than it took to compress it in the first place. As with reciprocating engines, the efficiency of the jet engine increases as the difference in temperature between combustion and exhaust increases. This means the higher the compression the more powerful and efficient the engine becomes. The difficulty of achieving high compression ratios without using a piston and cylinder meant it took several years development after their first success before jet engines finally displaced piston engines as the most efficient form of propulsion for aircraft. Building really good compressors is one of the most difficult jobs in engineering, and advances in compressor technology were slow and incremental.

sight. A technical name for a jet is a reaction propulsion engine. Reaction propulsion works in accordance with Newton's Third Law of Motion: for every action there is an equal and opposite reaction. This is the law which explains why lawn sprinklers and Catherine wheels spin round, and it also explains why guns recoil when they are discharged. The force that drives the bullet out of the barrel of the gun is opposed by an equal force acting to propel the gun backwards. It's only because the gun is so much heavier than the bullet that it doesn't accelerate so much. If the bullet was a lot heavier and the gun a lot lighter then you'd be better off holding on to the bullet and firing the gun at your enemy.

A gun could in principle be used as a reaction propulsion engine. Stand on a skateboard and fire a high-velocity machine gun behind you and the recoil acting on the gun would propel you around town at quite a lick – albeit at an unacceptable cost in human life. Dispense with the bullets and shoot hot gas out of the barrel instead and the recoil will still propel you around town, because the force of expansion that pushes gas out of the barrel will result in an equal and opposite force acting against the gun. This force is called thrust.

Whittle arrived at his idea for a turbo-jet by considering a scheme to use the exhaust from a piston engine to provide thrust to power an aeroplane – a recipro-jet, I suppose you could call it. This isn't quite as batty as it sounds. The massively powerful aero engines of the Second World War put out exhaust fumes with such a force that when they were directed backwards they provided a small but significant amount of thrust. But Whittle calculated that this type of jet would have little advantage over a piston-driven propeller plane: the fundamental problems faced by reciprocating engines in the upper atmosphere would be unresolved. A recipro-jet was designed and built, independently of Whittle's musings, by an Italian aircraft company, Caproni Campini. They constructed a reciprocating engine with a freewheeling crankshaft. Beyond overcoming their own friction and turning a flywheel the pistons didn't do any work. The engine's exhaust gases were channelled through the manifold and into a jet pipe. The force the expanding exhaust exerted on the jet pipe was sufficient to propel the aeroplane forward and allow it to take off, but the fuel consumption was excessively high and power output low.

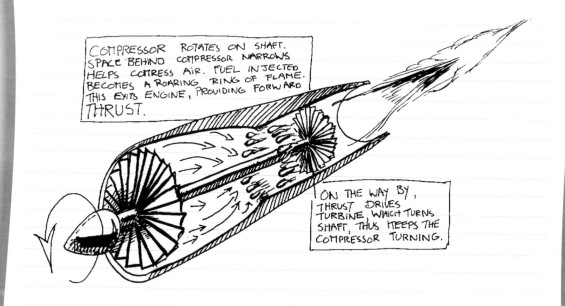

COMPRESSOR ROTATES ON SHAFT. SPACE BEHIND COMPRESSOR NARROWS HELPS COMPRESS AIR. FUEL INJECTED BECOMES A ROARING RING OF FLAME. THIS EXITS ENGINE, PROVIDING FORWARD THRUST.

ON THE WAY BY, THRUST DRIVES TURBINE WHICH TURNS SHAFT, THUS KEEPS THE COMPRESSOR TURNING.

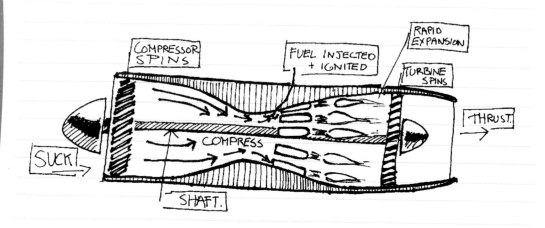

COMPRESSOR SPINS

FUEL INJECTED + IGNITED

RAPID EXPANSION

TURBINE SPINS

THRUST.

COMPRESS

SUCK!

SHAFT.

AXIAL-FLOW VS CENTRIFUGAL

Whittle's first patent specified an axial-flow compressor. Axial-flow compressors always promised to be – and eventually became – the most efficient type. But when it came to constructing a practical jet engine Whittle – like his unknown German co-worker, von Ohain – used a centrifugal compressor. Axial compressors were so inefficient at that time that dozens of them had to be used in series if adequately high pressures were to be attained – gas turbines that used the axial compressors of the day were heavy enough to snap the wings off most aircraft. But partly as a result of supercharger development, lightweight centrifugal compressors had reached a pitch of development far beyond that of the axial-flow compressor. Unfortunately, turbines had not reached a similar pitch of development and the power output of Whittle's early jets was severely limited by the ability of his turbines to withstand the constantly high temperatures of combustion.

"I don't have a tremendous affection for things that look Heath Robinson. I like things that look like they work really, really well. But the jet almost goes too far. Once you've seen inside, it becomes so simple it lacks all mystery. And it's footer free, which in one way is great and exactly what you want, but in another way it's a bit dull, because there's no interaction."

It was after rejecting the idea of a recipro-jet that Whittle looked into the possibility of using a gas turbine to provide jet propulsion instead. His calculations showed that it would do the job. He took his scheme to the Air Ministry to see if he could get official backing to do development work on his idea. But unfortunately for Whittle his scheme was referred to an official Air Ministry researcher, one Dr A. A. Griffith. Griffith was doing his own research on gas turbines, not to provide jet propulsion but to drive a conventional propeller – what is now known as a turbo-prop. He was struggling to get Air Ministry backing himself and it seems likely that his response to Whittle's proposal was influenced by his own interests. The Ministry wrote to Whittle and informed him that the materials necessary to build a gas turbine that was suited to the task of providing jet propulsion did not exist – which pretty much amounted to telling Whittle that his idea to build a jet engine was silly because jet engines hadn't been invented yet. Disappointed but not deterred, in January 1930 Whittle applied for a patent.

Apart from Griffith's unhelpful influence, Whittle also had a 1920 report on the suitability of the gas turbine as an aero engine by one Dr W. J. Stern quoted at him. So far this book has been filled with the stories of inspired technologists, people who were full of imaginative passion for the ways in which natural forces could be harnessed and

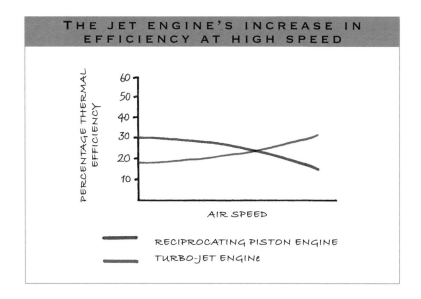

THE JET ENGINE'S INCREASE IN EFFICIENCY AT HIGH SPEED

PERCENTAGE THERMAL EFFICIENCY

AIR SPEED

⸺⸺⸺ RECIPROCATING PISTON ENGINE

⸺⸺⸺ TURBO-JET ENGINE

exploited for human ends. Dr Stern deserves mention as a foil to these great engineers. People sometimes say that engineering is an unimaginative business. It's true that number-crunching is central to modern engineering practice, and that most professional engineers are under-workers who apply the original thought of better minds. But that said, there's surely as much derivative and unimaginative work in the history of art and literature as there is in the history of engineering. And surely ground-breaking engineers use their imagination to transform reality just as much as good artists do, in the same way that good art, like good engineering, demands experimentation, precision and objectivity.

Dr Stern is a perfect example of an outstandingly unimaginative engineer. No doubt he was good at maths and his calculations were all correct. But getting your calculations correct is no use at all unless you're asking the right questions. Stern's calculations were hopelessly correct. Asked by the Air Ministry to report on the possibility of utilising the high potential power-to-weight ratio of the gas-turbine engine to provide a new kind of engine for aeroplanes, Dr Stern based all his calculations on the heavy, massively inefficient industrial gas turbines of the day. He worked out that, using cast iron for the combustion chamber, bronze for the turbine blades and an enormous multi-stage axial-flow compressor, the gas turbine would be far too

"Whittle said, 'Hey, I've got an idea for a jet engine!' And everyone said, 'Ignore him and he'll go away.' It was different in Germany. Von Ohain said, 'Ive got an idea for a jet engine!' and everyone said, 'Hey! Good idea! Let's give it a shot."

heavy to allow a plane to take off. Whittle performed the same type of calculations, but with a degree of creative optimism regarding what was possible.

This was the first in a series of painfully frustrating encounters with Government authorities that would have nipped the enthusiasm of a less pugnacious individual in the bud. Perhaps it was because Whittle did have great imagination that he and officialdom were never destined to see eye to eye. Whittle had a nasty habit of having new ideas and telling people about them. With a few honourable exceptions, civil servants seem to have found this habit distasteful in the extreme. And because Whittle was a serving RAF officer – and wished to remain in that condition – he was always obliged to get official permission for his activities.

Whittle's jet engine first took to the air in May 1941

Whittle failed to secure government backing for his plans, and he couldn't raise venture capital either. The situation was so bad that when his patent came up for renewal in 1935 he couldn't raise the necessary £5 fee. But shortly after the patent lapsed – when Whittle had almost given up hope of ever being given the opportunity to develop his idea – a sudden flurry of enthusiasm amongst a few interested individuals resulted in the formation of Power Jets, a company devoted to developing his plans. The Air Ministry, who have rights over any inventive work done by active servicemen, agreed to take a 25 per cent share in the company. Hilariously, they also decided that the work of Power Jets did not warrant official secrecy – because it was unlikely ever to be of military use. When war threatened they swiftly changed their minds, putting Whittle in a crazy Catch-22 situation. The Ministry refused to fund Whittle adequately on the grounds that if his jet engine was such a good idea, it should be easy enough for him to get city money for it. But because they had given Whittle's work the status of an official secret, he wasn't allowed to tell the backers he approached what his good idea was.

After his period as a marine test pilot Whittle had gone on to the RAF Officer's School of Engineering. He finished that course early and

Although the Meteor's engines pointed straight towards the future, its airframe construction was rooted firmly in the past. Its bodywork was pure 1940s, with lots of wires and wooden bits – more an MG than a Mazda. Accordingly, it has to be flown with a certain amount of respect.

subsequently had been allowed to go to Cambridge University where he studied Mechanical Sciences. When Power Jets was formed in 1936 he was in his final year at Cambridge, and despite devoting most of his time to practical work on his jet engine scheme still managed to get a First. The following year – on 12 April 1937 – Whittle started up the first ever turbo-jet. It was called the WU, which stood for Whittle Unit. The first run was a qualified success: the engine ran out of control; the combustion chamber glowed red hot and the Power Jets team ran for their lives – except for Whittle himself, who stayed and switched off the fuel supply. No one was hurt. The problem persisted for a while, but it turned out not to be a fault in the basic design; it was simply that fuel was leaking into the combustion chamber overnight.

Unlike the Air Ministry, the RAF was generous in its support of Whittle's efforts, putting him on a 'special duties' list and continuing to pay him an officer's salary while he worked on his jet. But despite his early success, funding for his enterprise was pathetically meagre. Whittle and his loyal team battled on and when war threatened, the Air Ministry, belatedly realising that they might have a war winner on their hands, ordered a flight engine from Power Jets and a plane from Glosters to put it in. The company staff was gradually augmented, funding problems eased and the development work increased in pace. On 15 May 1941, without ceremony or ado, the

"The speed and power of the jet engine totally changed the way wars were fought."

"The Meteor is fantastic. Because it's so small. It's the difference between being in a limo and a racing car. In a racing car everything's set hard, everything's very responsive and skittish. You feel every bump in the road. You have to to drive at that speed. But when you're in a big Cadillac everything's designed to help you deny the fact that you're actually travelling. And that's what a big airliner's like. You get in, you sit down, you shut the door, you put the television on, have a couple of drinks, chat the air hostess up and fall asleep. The whole conspiracy is to deny the fact that you're in a tin can three miles above the Earth. Whereas in a fighter there's no nonsense about it. There's the engine, there's the controls . . . you have the best imaginable view around you. You feel phenomenally exposed. It's like being in a tiny British sports car – a Triumph TR2. Your shoulders are actually above the level of the plane. You've got a perspex bubble over your head. Everywhere you look is sky – and when you're upside down, my God, I can't tell you! And of course we were flying low a lot of the time. You're doing 350 miles an hour and you could see people in their gardens!

And you know when you're on a plane and you hit an itsy little bit of turbulence and it drops a little and everyone spills a bit of tea on their lap – well this thing drops like a stone, then turns like it's being pulled round on the end of a rope, and your body's squeezed against the side and your face starts falling off your bones. It's fantastically exciting – a real sense of moving in three dimensions, which you don't get more than a hint of in a big plane. Every little twist is like somebody kicking you, like being tossed around on a bungy. It's so exciting. And the concentration needed to fly one of those things! You must have to clear your mind utterly of anything other than the business in hand."

first British jet-powered plane – the Gloster E. 28/39, powered by the Power Jets W1 engine – made its maiden flight.

Despite Power Jets' success, jet engine production was wrested from Whittle's control and vested in Rolls-Royce, whose version of Whittle's engine was called the Welland. The early Gloster Meteors were powered by Wellands. They were only a little faster than existing propeller planes, and the Air Force dared not use them over enemy territory lest one should be shot down and the 'secret' of the jet fall into enemy hands. The Germans had their own jet fighter, the Messerschmitt ME262, which they guarded with equal jealousy.

The Meteor was mainly used for shooting down German flying bombs. The V-bombs themselves were powered by a kind of non-turbine jet engine called a pulse jet. The pulse jet works by allowing air pressure to build inside the combustion chamber simply as a result of the forward motion of the aeroplane. When the pressure has built to an adequate level the combustion chamber is closed off and fuel is injected. Fuel is burnt and thrust produced until the pressure falls, when the combustion chamber is opened to the air once again and pressure is allowed to build. So the flying bombs followed a series of arcs in the sky as their jets 'pulsed' on and off.

Legend has it that the best technique for destroying V-bombs without blowing yourself up was to fly alongside the bomb, get your wing underneath its wing and then flip it over onto its back, whereupon it would go crashing to the ground. So long as you did it before the bomb got as far as London it counted as a success, though local farmers would probably have had something to say about this. As the result of some nasty shenanigans on the part of the Air Ministry and the big manufacturers, Whittle was comprehensively cut out of the further development of the jet, despite his brilliant early contribution and his continuing flow of excellent ideas – ideas which, had they been supported, would very likely have kept Britain ahead in the jet stakes for years and led to a more rapid development of jet power too. Over his life Whittle amassed drawers full of public honours, a bit of money from the Government and numerous academic awards, but his further help in developing the jet was actively resisted. Eventually he emigrated to the USA, where he lived for the rest of his

"If the jet engine had been lacking in character, then it was about to get it in spades. the Americans took Whittle's engine and added . . . sex."

The Pratt & Whitney J57 – the engine that took the world into the jet age. This one jet could develop thrust equivalent to 15 Spitfire engines.

"In Vietnam the jet engine allowed the bombers to operate from such a range and height that the planes were neither visible nor audible from the ground."

life. He remained hurt and baffled as to why his contribution to engineering in this country should have had such a hostile reception – and it is indeed a baffling matter.

In relation to other nations, the British Government's attitude to Whittle's invention can only be described as abject. In the same year that Whittle got his first jet off the ground his W1 jet engine was shipped to the USA. Fair enough, you might say, considering there was a war on. But for a one-off fee of only $800,000 the USA was given information on all of Whittle's projects, and the unlimited right to exploit Whittle's work for military and civilian uses, together with Power Jets' only bench-test engine and a support team. The General Electric company started development of the first US jet engine, the GE1, straight away, and by October 1942 the first US jet-propelled aircraft – the Bell XP59A – had made its maiden flight. Then in September 1946, when the war was over and the wartime coalition had been replaced by a Labour administration, the British Government made a grand gesture of friendship to the Soviet Union. They sent them a shipment of Rolls-Royce's best new jet engine, the Nene. The Russians had been struggling to do what they could with what they had retrieved from the gutted German turbo-jet factories and were as pleased as they were astonished by this gift. They copied the Rolls Royce Nene and went on to licence its production to China – for a fee, of course.

In the United States the Nene was licensed by Rolls-Royce to Pratt & Whitney, who replaced its centrifugal compressor with an axial-flow compressor and turned it into the next big thing in jet propulsion, the J57. The J57 powered the F100 – the first plane to break the sound barrier in level flight. It did so courtesy of a little gismo the Americans tacked on to the back of the engine – the afterburner. The axial-flow compressor that Pratt & Whitney developed for the J57 was so much more powerful than the centrifugal compressors of the early jets that it wasn't possible to burn enough fuel in the combustion chamber to use up all the oxygen in the air. So someone came up with the grand idea of burning even more fuel in the jet pipe. Press the red button and you can increase the thrust by over 50 per-cent almost instantaneously.

THE B52

Eight J57s – without afterburners – powered the early B52s, the most horribly successful bomber of the post-war years. At the height of the cold war, B52s flew the non-stop 'Chrome Dome' missions around the Arctic in readiness for a nuclear strike against the Soviets. The cruise missiles they carried in their holds were powered by a scaled-down version of the J57. Asked if the success of the US military jet program made America arrogant, an ex-B52 flight mechanic said 'No, it didn't make us arrogant. It just made the world aware that if necessary, America would destroy it.'

More than a quarter of a century after they were first introduced, B52s – powered by updated fan-jet engines – flew the majority of the bombing missions in the Gulf War.

The F100 Supersaver was introduced to the American airforce in 1957. On its first flight the pilot took it up to 35,000 feet, opened the throttle, and smashed through the sound barrier.

"This is where old airplanes come to die . . . the US military recycling center (AMARC) where the jets get turned into dogfood cans."

"The afterburner summed up America's attitude to the jet engine – pure raw power at any expense."

Other nations apart from the US picked up Britain's new technology and ran with it. Development went on in leaps and bounds. Axial-flow compressors became more and more powerful, turbines became able to withstand higher and higher temperatures, planes got bigger and/or faster and were able to fly higher and higher. But the development of the jet did not continue along the same trajectory for long. Though for a while it looked as if the propeller had become a thing of the past, it soon made a comeback in the modified form of the turbo-fan. Whittle had forseen as much – indeed, he had patented the idea back in 1936 – but had not been given the facilities to develop it.

"Being that close to an afterburner was a wee bit like hell. I remember wondering, Is this what hell is meant to be like? The power of it is so far beyond anything we can get hold of. You suddenly realise what a vulnerable thing a human being is. We're just these wee monkeys clinging on a rock. Then when I put my marshmallow in it I thought, Okay: now toast my marshmallow, big boy!"

So it wasn't until the early seventies, when fuel prices started to go up, that the turbo-fan saw substantial development.

The fan is the bit you see at the front of jet engines. Today even fast fighter-planes have fans. The fan is a many-bladed propeller, with the added advantage that airflow is carefully guided around it by the engine cowling. The fan drives some air into the engine itself – acting as a first stage compressor – and around the outside of the engine too. The ratio of the air that passes outside the engine to the air that passes through the engine is called the engine's bypass ratio. The first turbo-fan engines had very small bypass ratios, but bypass ratios on today's big civil jets have gone up to eight to one, and are still increasing. Turbo-fans are quieter, cleaner and more fuel efficient than the straight-through or 'stovepipe' jets that preceded them.

"People have an immense capacity to reconcile themselves to technological change. When the jet went into passenger planes people were scared because they couldn't see the propeller going round. How could a plane fly without a propeller? But I was in an ATP recently – an advanced turbo-prop – and this woman sitting beside me was really upset. She said 'Look at that big propeller going round. It's really scary isn't it!'"

The reason for this increased efficiency is simple. As we have seen, Jet engines, like propellers, work by accelerating a mass of air backwards. In accordance with Newton's Third Law of Motion, the force that is used to accelerate the air backwards is equalled and opposed by a force acting on the engine or propeller. This is the forward thrust that the jet or the propeller produces. But the same thrust can be generated by accelerating a great mass of air to a low speed or by accelerating a small mass of air to a high speed. Which route you choose to take makes no difference as far as the amount of thrust you create is concerned, but it makes an enormous difference when it comes to the amount of fuel you consume in creating it. Because the fuel energy that is swallowed up in creating that thrust is directly proportional to the kinetic energy imparted to the mass of gas. The kinetic energy of the gas is equal to $\frac{1}{2}MV^2$ – that is, half the product of the mass of the gas and the square of the velocity. This means kinetic energy increases numerically as the mass increases, but exponentially as the velocity increases. Accelerating a large mass of gas to a low velocity takes more energy than accelerating a small mass of gas to a high velocity, and so it takes more fuel. Of course, if the plane you are powering is moving at more than twice the speed of sound it is necessary to accelerate the air to extremely high velocities in order to generate any thrust at all. But if your plane is travelling only at speeds of 500 or 600 m.p.h. it is much better value to accelerate large masses of air to relatively low velocities. So – paradoxically – the cutting edge work in jet engine technology is now directed towards the development of efficient high-speed propellers.

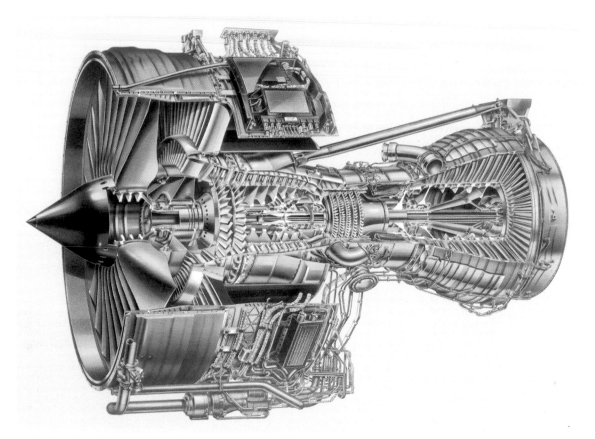

I know I said jet engines were simple. . . but the principle is the same, honest. The Rolls-Royce Trent is pretty much state of the art. The most powerful version is a hundred times more powerful than Whittle's first turbo-jet: four of these lightweight monsters will generate enough electricity to power a city the size of Derby. Their ultra-efficient two-stage axial compressor gives a pressure ratio in the combustion chamber of 40 to 1 – extracting power from paraffin with an efficiency Rudolf Diesel would have drooled over.

Although the design is sophisticated, at the end of the day the bounds of the possible in engineering still comes down to metals technology – just as in James Watt's day. To take one example, each of those delicate-looking blades in the nine foot diameter fan is subject to a load of 95 tons – the weight of ten double-decker buses.

The jet engine was a true stroke of genius. It has everything you could ask of an engine: power; economy; reliability; low size and weight. It isn't quite in the class of a Cadillac V8 or a supercharged Bentley as far as the noise it makes is concerned, but you can warm to that high-pitched whine. The impact it has had on long-distance travel is truly immense. The jet has done on a global scale what the steam engine did on a national scale, bringing the possibility of economical travel to millions and millions of people. And perhaps the most impressive thing of all is that the jet achieved this not through the complexity but through the simplicity of its design.

"I think we can conclude that the jet is a good thing when it's taking you to Marbella to get pissed for a fortnight, and a bad thing when it's carrying bombs to be dropped on people. But thinking about it . . . if you live in Marbella it amounts to pretty much the same thing, doesn't it?"

APPENDIX

LOCATIONS

V8

FORD GALAXY
Highlands west of Loch Lomond, Scotland

FORD FLATHEAD ROADSTER
Boulder Oaks, near San Diego, southern California

EDELBROCK HOT ROD
New Smyrna Beach, Florida

CHEVY BEL AIR
Tuscon, Arizona

CADILLAC ELDORADO
near Lancaster, Mojave Desert, California

NASCAR
Finish Line Racing School, near Daytona, Florida

STEAM

FIREQUEEN
Loch Lomond, Scotland

SS SHIELDHALL
The Solent, Hampshire coast, England

NORTH BRITISH 12TH CLASS
Victoria Falls Bridge, Zambezi River, Zimbabwe

MAINTENANCE WORK
The National Railways of Zimbabwe Steam Shed, Bulawayo, Zimbabwe

BEYER GARRATT 15TH CLASS
The Hwange Colliery, Hwange Town, Zimbabwe

TWO-STROKE

TRABANT
Chemnitz, eastern Germany

RADIO-CONTROLLED MODELS, JET-SKIS, TRAIL BIKES
Loch Lomond, Scotland

MZ FACTORY AND BIKES
Zschopau, eastern Germany

CHAINSAW, SNOWMOBILE
McCloud, northern California

VESPA
Trieste, Italy

DIESEL

DIESEL MUSEUM, MAN
Augsburg, southern Germany

LIZARD LIGHTHOUSE
Lizard Point, Cornwall, England

CATERPILLAR TRACTOR
Sacramento, northern California

CONTAINER SHIP
en route from Rotterdam, Netherlands to Southampton, England

TRUCK AND TRAIN
Sacramento, northern California

SULZER FACTORY
Trieste, Italy

SUPERCHARGER

MERCEDES-BENZ
Nürburgring, southern Germany

BENTLEY
Brooklands, Weybridge, Surrey, England

SPITFIRE AND MESSERSCHMITT
Duxford Air Museum, Cambridgeshire, England

JET

AVRO 504K
*Shuttleworth Collection, Old Warden Airfield,
Cambridgeshire, England*

GLOSTER METEOR
Jet Heritage, Bournemouth Airport, Hampshire, England

B52 GRAVEYARD
*American Military Aircraft Recylcing Center (AMARC),
Tucson, Arizona*

F100 AFTERBURNER
Lancaster Airport, Mojave Desert, California

**VIRGIN ATLANTIC TRAINING
FACILITY**
Heathrow Airport, Middlesex, England

PICTURE CREDITS

John Binias

Chris Chapman

John Francas

Murdo McLeod

John Rogers / Frank Spooner Pictures

The Mitchell Library, Glasgow

New Sulzer Diesel Ltd

MAN Aktiengesellschaft Historiches Archiv/Museum

Rolls-Royce plc

Every effort has been made to trace the copyright
holders of the photographs used in this book, but one
or two remain unidentified. We would be grateful if
the photographers concerned would contact us.